Elements

OXYGEN

O

Atlantic Europe Publishing

How to use this book

This book has been carefully developed to help you understand the chemistry of the elements. In it you will find a systematic and comprehensive coverage of the basic qualities of each element. Each two-page entry contains information at various levels of technical content and language, along with definitions of useful technical terms, as shown in the thumbnail diagram to the right. There is a comprehensive glossary of technical terms at the back of the book, along with an extensive index, key facts, an explanation of the Periodic Table, and a description of how to interpret chemical equations.

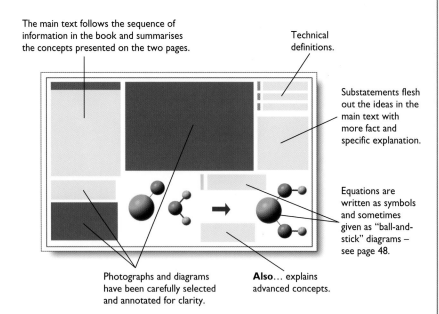

The main text follows the sequence of information in the book and summarises the concepts presented on the two pages.

Technical definitions.

Substatements flesh out the ideas in the main text with more fact and specific explanation.

Equations are written as symbols and sometimes given as "ball-and-stick" diagrams – see page 48.

Photographs and diagrams have been carefully selected and annotated for clarity.

Also… explains advanced concepts.

- -

An Atlantic Europe Publishing Book

Author
Brian Knapp, BSc, PhD
Project consultant
Keith B. Walshaw, MA, BSc, DPhil
        (Head of Chemistry, Leighton Park School)
Industrial consultant
Jack Brettle, BSc, PhD (Chief Research Scientist, Pilkington plc)
Art Director
Duncan McCrae, BSc
Editor
Elizabeth Walker, BA
Special photography
Ian Gledhill
Illustrations
David Woodroffe
Electronic page make-up
Julie James Graphic Design
Designed and produced by
EARTHSCAPE EDITIONS
Print consultants
Landmark Production Consultants Ltd
Reproduced by
Leo Reprographics
Printed and bound by
Paramount Printing Company Ltd

Suggested cataloguing location
Knapp, Brian
        Oxygen
        ISBN 1 869860 54 3
        – (*Elements* series)
540

Acknowledgements
The publishers would like to thank the following for their kind help and advice: *The Copper Development Association.*

Picture credits
All photographs are from the **Earthscape Editions** photolibrary except the following:
(c=centre t=top b=bottom l=left r=right)
courtesy of **The Copper Development Association** 28bl;
Catherine Gledhill 39bl; **NASA** 6/7tl, 39tr and **ZEFA** 7cr, 8br, 29.

Front cover: As the carbon compounds that make up the plant tissues in a forest fire are oxidised, they produce a strongly exothermic reaction and release carbon dioxide gas. The carbon dioxide released by burning forests is an important contributor to global warming through the Greenhouse Effect.
Title page: Almost all the oxygen present in the atmosphere of the Earth was and is still formed by green plants and is a byproduct of photosynthesis.

First published in 1996 by
Atlantic Europe Publishing Company Limited, Greys Court Farm,
Greys Court, Henley-on-Thames, Oxon, RG9 4PG, UK.

Copyright © 1996
Atlantic Europe Publishing Company Limited
First reprint 1997   Second reprint 1997

This product is manufactured from sustainable managed forests. For every tree cut down at least one more is planted.

The demonstrations described or illustrated in this book are not for replication. The Publisher cannot accept any responsibility for any accidents or injuries that may result from conducting the experiments described or illustrated in this book.

Contents

Introduction

An element is a substance that cannot not be decomposed into a simpler substance by any known means. Each of the 92 naturally occurring elements is therefore one of the fundamental materials from which everything in the Universe is made. This book is about oxygen.

Oxygen

Oxygen is the most common element by volume or mass (weight) on Earth. In each breath you take, one-fifth of the molecules are oxygen; there are twelve trillion tonnes of oxygen in the air. Yet despite the fact that it is so common, we are hardly aware of this vital element because oxygen is transparent, it has no colour, taste or smell.

Oxygen occurs uncombined (as one of a mixture of gases) in air, but it also readily combines (reacts) with a wide variety of other elements to make compounds. Indeed, oxygen is among the most reactive of all the elements. For example, nearly all the rocks of the Earth are compounds containing oxygen. Water is also a compound of oxygen along with hydrogen.

Oxygen is vital for life. When oxygen reacts with a fuel, the fuel is oxidised and becomes an oxide, sometimes turning into a solid, in other cases forming a gas such as carbon dioxide. This reaction – called burning – releases heat energy.

The reaction of oxygen with fuels such as coal, oil and natural gas can also be harnessed to convert the chemical energy in the fuel into movement energy. This is how we power many of our machines.

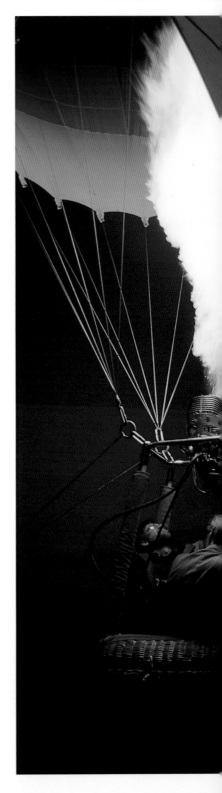

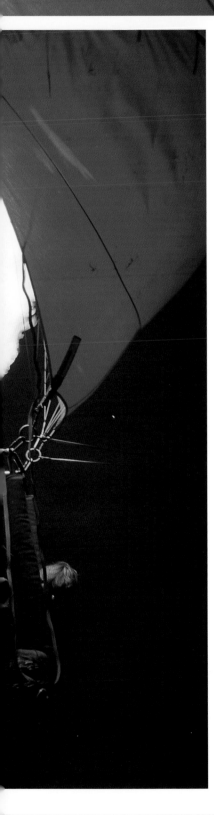

All kinds of fuels can "burn" in the sense of being oxidised by oxygen, although not all give out large amounts of heat. The carbohydrates in our food, for example, oxidise to release energy for our bodies. When dead organic material decays, it is actually oxidising. Oxidation also happens to food left exposed to the air and to the cells in our bodies.

Oxygen reacts with most metals. Sometimes the effects cannot be seen, such as when an invisible oxide film forms on the surface of a sheet of aluminium. But other metals react more dramatically. Iron, for example, reacts with oxygen in the presence of water to produce the familiar flakes of rust, while potassium instantly catches fire.

The formula for oxygen gas is written as O_2. This means that oxygen gas (like most other gases) normally occurs as a molecule, made of two atoms, rather than as a single atom. Most single atoms of oxygen occur high in the atmosphere. Oxygen can also occur as a molecule with three atoms, in which case the chemical symbol is O_3. This form of oxygen is called ozone and it has quite different properties from the more plentiful O_2, including the property of shielding living things from the harmful ultraviolet radiation of the Sun. Scientists are worried that the amount of ozone in the atmosphere is now so low that it is no longer an effective shield.

◀ This flame is created by the reaction of propane gas and oxygen. The heat given off by the reaction is quite sufficient to warm the air and lift the balloon. (The other products of the combustion reaction are carbon dioxide gas and water vapour.)

Oxygen in the air

Air is a gas made up chiefly of nitrogen and oxygen. Oxygen (O_2) makes up 21% of the atmosphere by volume and 23% by mass (weight). The weight of oxygen molecules in the atmosphere helps to create the air pressure on the Earth's surface, on average one kilogram on every square centimetre.

Although ozone (O_3) is only present in concentrations as little as 12 parts per million, it is present both in the upper atmosphere (see page 12) and close to the ground (see page 42).

How the atmosphere gained its oxygen

The gases in the early atmosphere (over three billion years ago) were produced during volcanic eruptions. The gases would have included water vapour and carbon dioxide, but there was no free oxygen. The first free oxygen must have been created from these compounds.

The oldest rocks certainly formed in an oxygen-poor atmosphere. When exposed to the air, these very old rocks oxidise (react with oxygen) rapidly.

Oxygen was probably produced at first by the effects of light energy on water. This would have released free hydrogen and oxygen. Yet this process could only have produced about 1% of the oxygen now in the atmosphere. The rest must have been produced in a different way.

The lack of oxygen in the early atmosphere also means there was no ozone in the upper atmosphere to shield life from the harmful effects of ultraviolet light, so this may be why the first life forms developed in the protected surface layers of the oceans, rather than on land.

Most scientists believe that the vast majority of the oxygen now in the atmosphere was produced in the last three billion years of the Earth's history, when ocean plants began to release oxygen as part of the process called photosynthesis (see page 14). Oxygen must gradually have seeped out of the water and into the atmosphere. Eventually the build-up of oxygen in the air provided the conditions suited to life on the surface of the Earth.

❶ The accumulation of oxygen created the present mixture of gases found in the modern atmosphere.

❷ The oxygen that could not be absorbed in the ocean waters seeped into the atmosphere.

❸ The earliest life forms were plants in the oceans. They released oxygen as a waste product.

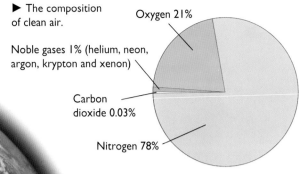

▶ The composition of clean air.

Oxygen 21%

Noble gases 1% (helium, neon, argon, krypton and xenon)

Carbon dioxide 0.03%

Nitrogen 78%

aurora: the "northern lights" and "southern lights" that show as coloured bands of light in the night sky at high latitudes. They are associated with the way cosmic rays interact with oxygen and nitrogen in the air.

molecule: a group of two or more atoms held together by chemical bonds.

oxidise: the process of gaining oxygen. This can be part of a controlled chemical reaction, or it can be the result of exposing a substance to the air, where oxidation (a form of corrosion) will occur slowly, perhaps over months or years.

▲ An aurora forms a beautiful veil shape in the sky that is clearly visible at night.

Also...

Oxygen is less common than hydrogen, helium, and neon in the Universe, but by far the most common element on the Earth. It makes 90% of water (by mass), about 47% of the Earth's crust, as well as 23% of dry air and 65% of the human body.

◀ Almost all the oxygen present in the atmosphere of the Earth was and is still formed by green plants as a byproduct of photosynthesis, which involves the conversion of carbon dioxide and water to glucose.

Auroras

Auroras are bands of light that appear in the night sky of high-latitude regions. The bands occur when cosmic rays reach the outer atmosphere, and in particular interact with the oxygen molecules. The result is the production of crimson and greenish-white bands that form shapes like curtains moving across the sky.

The pink and violet bands are caused by the effect of cosmic rays on the nitrogen in the atmosphere, but these are less prominent than the oxygen bands.

Oxygen in water

In the past, oxygen atoms combined with hydrogen atoms to form water. Oxygen is a much more massive element than hydrogen (the atomic weight of hydrogen is 1, that of oxygen is 16). As a result, oxygen makes up about nine-tenths of the mass of the oceans.

The oxygen in the compound water is very strongly bonded to the hydrogen atoms and it will not break up even on heating. This is why when water boils the molecules of water change from liquid water to water vapour, rather than breaking up into hydrogen and oxygen atoms.

The modern atmosphere and oceans contain almost no free hydrogen, so oxygen and hydrogen can no longer combine. Because there is a surplus of oxygen atoms, the atmosphere contains free oxygen. Oxygen also occurs as free oxygen molecules dissolved in water. These molecules do not combine with the water molecules. In most surface water there may be 45 grams of oxygen dissolved in every cubic metre of water. These are the oxygen molecules that are vital for life in the water.

▲ Stagnant water is still water with little oxygen in it. This situation occurs when microbes living in the water use the oxygen faster than the rate at which oxygen can dissolve into the water. Because oxygen dissolves only slowly in water, still water has a low oxygen content. In warm conditions, microbes grow quickly and can soon use up the dissolved water making it stagnant. This is why stagnant conditions are more common in summer than winter.

▶ One of the key processes in a water treatment plant is aeration. Artificial fountains aerate the water, providing the oxygen needed by the bacteria that consume waste organic material.

◄▼ Oxygen dissolves poorly in still water. It has to be mixed into the water physically. In the open oceans this mixing takes place through the action of waves. Lakes become aerated by the action of surface waves. Thus, oxygen is most abundant in the upper levels of any body of water and this, therefore, is also where most life is found.

In rivers oxygen is mixed with the water as it tumbles over rocks, goes over waterfalls or flows in river channels.

dissolve: to break down a substance in a solution without a resultant reaction.

ionise: to break up neutral molecules into oppositely charged ions or to convert atoms into ions by the loss of electrons.

ionisation: a process that creates ions.

molecule: a group of two or more atoms held together by chemical bonds.

vapour: the gaseous form of a substance that is normally a liquid. For example, water vapour is the gaseous form of liquid water.

Also...

The chemical symbol for water is normally written H_2O, meaning that water molecules contain three atoms: two hydrogen atoms bound to each oxygen atom. The bonds are very strong and rarely break. However, a few water molecules ionise, that is the neutral molecules break up into oppositely charged particles – hydrogen ($H^+(aq)$) ions and hydroxide ($OH^-(aq)$) ions.

The ions in a liquid allow it to behave as an electrolyte – a carrier of electric current. Because water ionises so poorly it is a poor conductor of electricity. This is why, for example, pure water cannot be used alone in a vehicle battery. Compounds that readily ionise (substances like dilute sulphuric acid) have to be added to improve the conductivity.

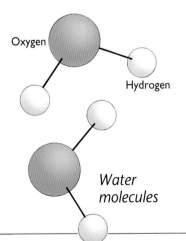

Oxygen

Hydrogen

Water molecules

This part of the diagram represents a hydrogen ion associated with a water molecule.

This part of the diagram represents a hydroxide ion.

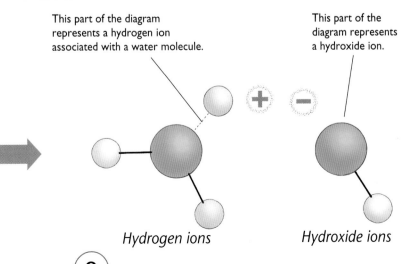

Hydrogen ions

Hydroxide ions

Oxygen in minerals and rocks

Oxygen makes up about half of the mass (weight) of material in the Earth's crust,

Most minerals are compounds of oxygen and the majority of these are silicates or oxides. In all silicates, more than half the atoms are oxygen. Together, silicon and oxygen comprise three-quarters of the mass of the Earth's crust. This means that when you look at any soil or rock you are probably looking at a solid, locked-together mass of silicon and oxygen.

Silica

Silica is a molecule of one silicon atom and two oxygen atoms locked together by strong bonds. This unit forms the building block for many of the world's minerals. In its pure form it makes up silicon dioxide, quartz, also known as silica, and the minerals that it forms are called silicates. It takes 18 million of these tiny units to make a piece of silicate 1 centimetre long.

Many of the world's gemstones – such as sapphire, aquamarine and emerald – are silicates, combining metal ions with silica.

All the familiar building materials – glass, brick, pottery clay and even concrete – are also combinations containing oxygen and silicon.

▶ The structure of silica is very similar to diamond, with the silicon atoms packed tightly together. Four large oxygen atoms lie at the corners of a tetrahedron, almost entirely surrounding a smaller silicon atom. The electric charges between the silicon and the oxygen atoms balance exactly, forming strong bonds. This structure makes, among other minerals, quartz, which is very stable and unreactive.

Oxygen

Silicon

This is a model of part of a silica molecule. It is in the form of a four-cornered solid called a tetrahedron. This is the basic building block of most of the world's minerals.

Topaz

Topaz is a form of a silicate containing the metals aluminium and fluorine. It varies in colour and may be colourless, white, grey, yellow, orange, brown, light blue, light green, purple or pink. It is regarded as a gemstone.

Topaz crystal

Ruby set in a green groundmass of smaller crystals

gemstone: a wide range of minerals valued by people, both as crystals (such as emerald) and as decorative stones (such as agate). There is no single chemical formula for a gemstone.

ion: an atom, or group of atoms, that has gained or lost one or more electrons and so developed an electrical charge.

mineral: a solid substance made of just one element or chemical compound. Calcite is a mineral because it consists only of calcium carbonate, halite is a mineral because it contains only sodium chloride, quartz is a mineral because it consists of only silicon dioxide.

silicate: a compound containing silicon and oxygen (known as silica).

Ruby (aluminium oxide)

Ruby is a deep red crystal, one of the most prized of all gemstones. It is made mainly of aluminium and oxygen (aluminium oxide). This mineral, known as corundum, is transparent. However, when it occurs with small amounts of another element, chromium, the colour changes to somewhere between pale rose and deep red.

Haematite (iron oxide)

This is the name for the most widespread form of iron ore. It can be found as concretions (lumps), but it is most often found in rocks once deposited by rivers or the sea. Deep red rocks often tell of the iron oxide colour of haematite. Most such rocks were formed in those parts of the tropics with a wet and a dry season. During the wet season the minerals were eroded from the land and washed to inland basins, deltas or coasts. During the dry period the water evaporated and the sediments dried out, oxidising iron compounds to iron oxide. These rocks are often referred to as "red beds" and are usually fine-grained materials such as shales.

Haematite amongst quartz crystals

Ozone

Ozone (O_3), named for the Greek word for "smell" is a poisonous, colourless and tasteless gas with a distinctive strong smell. Molecules of ozone are probably the source of the smell that can be detected close to working electrical equipment such as motors and TVs. If a vehicle with a catalytic converter is started cold, ozone can be detected in the exhaust fumes.

The ozone layer

Most ozone is found high in the atmosphere, in a region of the stratosphere called the ozone layer. Here ozone performs a vital life-protecting role, absorbing the ultraviolet rays of the Sun that would be harmful to both plants and animal life.

The number of ozone molecules at this level is very small. If brought down to the ground surface they would form a gas layer no more than 3 mm thick. This is why the ozone layer can be so easily disrupted by human activities.

EQUATION: Reaction of CFCs with ozone

❶ *CFC + ultraviolet light ⇨ chlorine*

$$CFC(g) + UV\ light \Rightarrow Cl(g)$$

❷ *Chlorine atom + ozone ⇨ chlorine oxide + oxygen*

$$Cl(g) + O_3(g) \Rightarrow ClO(g) + O_2(g)$$

❸ *Chlorine oxide + oxygen atom ⇨ chlorine atom + oxygen*

$$ClO(g) + O(g) \Rightarrow Cl(g) + O_2(g)$$

◀▼ When chlorofluorocarbons (CFCs) reach the stratosphere they are broken down by ultraviolet solar radiation releasing chlorine atoms. These atoms attack the ozone molecules in the air, but the chain reaction that occurs releases a chlorine atom at the end. Thus a single chlorine atom can survive in the upper atmosphere for four to ten years. During that time it can destroy countless ozone molecules.

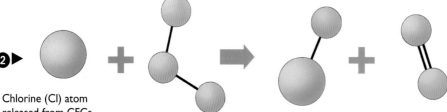

❷▶

Chlorine (Cl) atom released from CFCs and carried into the upper atmosphere by air currents.

Ozone (O_3) molecules are formed in the upper atmosphere.

This very reactive chlorine atom is called a "free radical". It can now react with another ozone molecule, causing further destruction of the ozone layer.

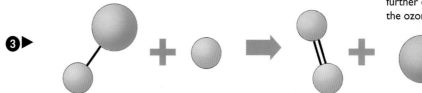

❸▶

Destruction of ozone in the stratosphere

Ozone mostly occurs in the layer of the upper atmosphere between 20 and 25 km above the Earth's surface. Here a natural balance between the plentiful oxygen (O_2) and ozone (O_3) molecules has existed for billions of years.

Ozone is produced by the effect of cosmic rays on oxygen (O_2) molecules and by electrical discharges, such as lightning flashes, working on (O_2) oxygen molecules. The energy from either source decomposes some oxygen (O_2) molecules, leaving the free atoms (O) to link to nearby oxygen molecules to make ozone (O_3).

Ozone is also decomposed by ultraviolet sunlight to release . oxygen both as molecules (O_2) and as atoms (O).

Ozone, however, does not just decompose in ultraviolet light. It can react with other molecules in the upper air, for example, chlorine and chlorine compounds.

The most troublesome of the chlorine-based compounds is a group known as CFCs. They have been widely used in aerosol spray cans, refrigerators, plastic foam and cleaning fluids. Chlorine containing molecules break up the ozone, taking away an oxygen atom and leaving O_2. Because this process is *in addition to* the natural decomposing reactions in the upper air, ozone is decomposing faster than it is forming.

As CFCs cause the ozone concentration to fall, the amount of ultraviolet radiation reaching the Earth rises, and the occurrence of skin cancers increases.

catalyst: a substance that speeds up a chemical reaction but remains unaltered at the end of the reaction.

decompose: to break down a substance (for example by heat or with the aid of a catalyst) into simpler components. In such a chemical reaction only one substance is involved.

free radical: a very reactive atom or group with a "spare" electron.

ion: an atom, or group of atoms, that has gained or lost one or more electrons and so developed an electrical charge.

▲ Ground level ozone cannot be seen, except in the smarting eyes of people walking along the streets.

◄ Traffic in Bangkok, Thailand. Along with all the other pollutants such as smoke particles, nitrogen dioxide, sulphur dioxide and carbon monoxide, ozone is also produced by car engines. Ozone has a distinct and pungent odour.

Ground level ozone

Ozone can also be produced as a result of city traffic. The poisonous properties of ozone become apparent here. Even at very low concentrations, ozone can cause eye irritation, and for those people suffering from asthma and other breathing disorders, places with high street-level ozone can cause great discomfort.

Ozone can also be harmful to the growth of plants, contributing to the stress on trees that may make them more liable to suffer the effects of acid rain.

The oxygen cycle

There is a continual exchange of oxygen between the atmosphere and the water, the plants and animals and mineral matter. This is called the oxygen cycle.

It begins with the reservoir of carbon dioxide in the air. The process of photosynthesis uses carbon dioxide and water from the soil to produce cellulose, the material from which plants are made. This releases oxygen gas into the air. Photosynthesis ceases when darkness falls and plants then burn off some of the cellulose they have made, returning carbon dioxide to the atmosphere.

Animals use oxygen in the atmosphere for respiration, oxidising the sugars in their food to give energy and releasing carbon dioxide to the atmosphere.

When dead tissue (carbon compounds) decays by a combination of oxidation and microorganism decay, carbon dioxide is released back to the atmosphere.

A slower cycle occurs whenever mineral matter is oxidised, such as in the formation of rocks.

▲ The decay of leaves is mainly an oxidation process.

Oxygen decomposes organic matter

If you leave vegetables out on a window ledge, they will eventually shrivel and then rot. This process is called decomposing.

Decomposing is a chemical reaction similar to combustion. The organic material of waste food, dead weeds and so on, will react with the oxygen in the air and change chemically. Because all this happens at normal temperatures, the reaction is slow. In many cases microorganisms also play an important role, such as in helping to rot down dead organic material. When butter or milk for example, taste unpleasant, it is because they are oxidising, often with the help of unseen microorganisms.

A compost heap is made of decomposing vegetable material. The compost is also a good insulator, which means that it keeps the heat inside the compost. If you feel inside a compost heap it is likely that it will be warm, if not hot. This shows that the organic material of the compost is oxidising and forming new materials and at the same time it is releasing heat. This shows that oxidation is an exothermic process.

EQUATION: Photosynthesis

Carbon dioxide + water ⇨ glucose + oxygen

$6CO_2(g) + 6H_2O(l) \Rightarrow C_6H_{12}O_6(s) + 6O_2(g)$

EQUATION: Oxidation of organic matter

Glucose + oxygen ⇨ carbon dioxide + water

$C_6H_{12}O_6(s) + 6O_2(g) \Rightarrow 6CO_2(g) + 6H_2O(l)$

combustion: the special case of oxidisation of a substance where a considerable amount of heat and usually light are given out. Combustion is often referred to as "burning".

exothermic reaction: a reaction that gives heat to the surroundings. Many oxidation reactions, for example, give out heat.

photosynthesis: the process by which plants use the energy of the Sun to make the compounds they need for life. In photosynthesis, six molecules of carbon dioxide from the air combine with six molecules of water, forming one molecule of glucose (sugar) and releasing six molecules of oxygen back into the atmosphere.

▼ The oxygen cycle.

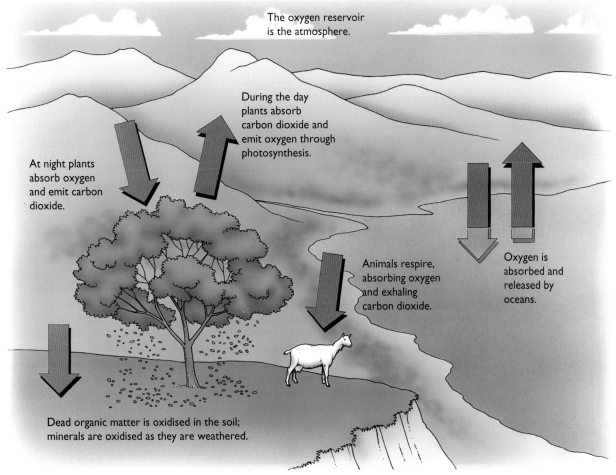

The oxygen reservoir is the atmosphere.

During the day plants absorb carbon dioxide and emit oxygen through photosynthesis.

At night plants absorb oxygen and emit carbon dioxide.

Animals respire, absorbing oxygen and exhaling carbon dioxide.

Oxygen is absorbed and released by oceans.

Dead organic matter is oxidised in the soil; minerals are oxidised as they are weathered.

Also...

The blood performs a number of essential functions, but one of the most important is to supply oxygen to the body. As the blood is pumped by the heart to the lungs, the blood passes close to the surface of the lungs, allowing oxygen in the lungs to diffuse into the blood. This process is part of respiration. The enriched blood then goes back to the heart where it is pumped into the arteries and around the body.

Blood transports oxygen by means of haemoglobin, an iron-based chemical substance present in all blood. Haemoglobin contains iron atoms to which oxygen temporarily bonds during transport to the tissues. The blood is capable of absorbing up to 25 millilitres of oxygen for every 100 millilitres of blood.

Making oxygen chemically

Pure oxygen has a large number of uses in the modern world. It is used, for example, in medicine, but also on a much larger scale for making fuels burn more effectively, such as in an iron-making blast furnace or in an oxyacetylene welding and cutting lance.

On a large industrial scale, atmospheric air is cooled. At -183°C oxygen liquefies; it can then be stored in tanks and transported easily.

Laboratory preparation

In the school laboratory oxygen is made by reacting two chemicals together or by causing an oxygen-rich compound to decompose.

It is common to use the apparatus shown on this page for small samples. Here an oxygen-rich compound, hydrogen peroxide in solution in water is dripped on to manganese oxide. The manganese oxide causes the hydrogen peroxide to decompose and release oxygen and water. The oxygen is collected in a gas jar over water using a support called a beehive shelf.

An alternative procedure involves heating a compound (for example potassium chlorate) which liberates oxygen as it is heated (i.e. as it decomposes).

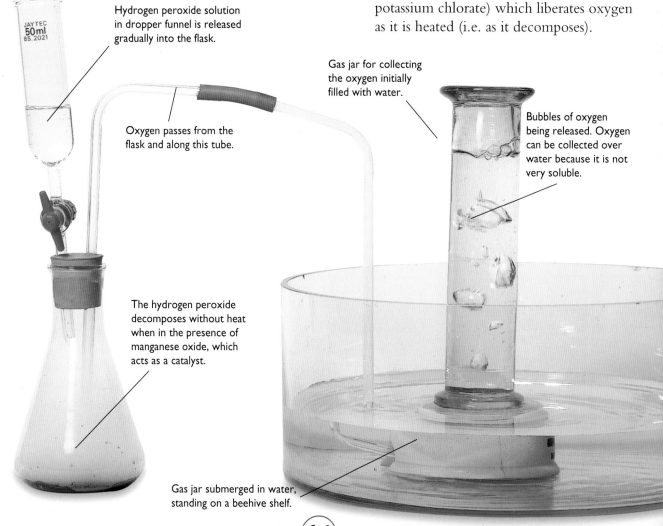

Hydrogen peroxide solution in dropper funnel is released gradually into the flask.

JAYTEC 50ml BS.2021

Oxygen passes from the flask and along this tube.

The hydrogen peroxide decomposes without heat when in the presence of manganese oxide, which acts as a catalyst.

Gas jar for collecting the oxygen initially filled with water.

Bubbles of oxygen being released. Oxygen can be collected over water because it is not very soluble.

Gas jar submerged in water, standing on a beehive shelf.

Industrial preparation of oxygen

The easiest way to obtain oxygen in large quantities is to liquefy air in bulk. This process is called distillation, and it is a very low temperature version of, for example, how oil and other mixtures are obtained. The process takes place in a tall column (called a fractionation column) made of a set of "leaky trays". Air is cooled until it becomes liquid and is then poured onto the trays. There is a boiler at the bottom and a condenser (a trap for the gas given off) at the top.

In this gas, the air begins to evaporate and nitrogen is given off as a gas, while the oxygen (which goes into a gas less easily than nitrogen – it is less volatile) drips to the bottom. In this way carefully controlled conditions can separate out the oxygen from other gases.

The liquid air is kept under a pressure of five atmospheres in the tower, at which pressure the boiling point of the nitrogen and oxygen are much higher (and so easier to maintain) than at normal atmospheric pressure.

catalyst: a substance that speeds up a chemical reaction but itself remains unaltered at the end of the reaction.

decompose: to break down a substance (for example by heat or with the aid of a catalyst) into simpler components. In such a chemical reaction only one substance is involved.

Nitrogen gas. Nitrogen boils at -196°C. It is used to make fertilisers, and nitric acid.

▶ Fractional distillation of air. The liquid air is pumped into a fractionating column and subjected to pressure and temperature changes to separate the component gases that make up air.

Argon gas. Argon boils at -186°C and is used as an unreactive filling for some light bulbs.

Air is cooled to -190°C and pressurised to generate liquid air.

Liquid oxygen. Oxygen boils at -183°C and is used for breathing apparatus.

EQUATION 1: Production of oxygen using hydrogen peroxide

Hydrogen peroxide ⇨ oxygen + water

$$2H_2O_2(aq) \xrightarrow{\text{catalyst}} O_2(g) + 2H_2O(l)$$

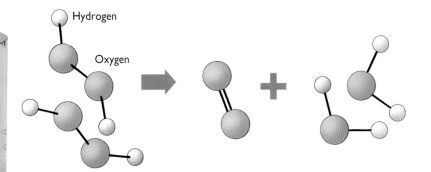

Hydrogen

Oxygen

EQUATION 2: Using potassium chlorate to make oxygen

Potassium chlorate ⇨ potassium chloride + oxygen

$$2KClO_3(s) \implies 2KCl(s) + 3O_2(g)$$

Also...

Manganese oxide acts as a catalyst, a compound that is essential to allow a reaction to happen quickly, but that actually does not change chemically itself.

The decomposition of potassium chlorate ($KClO_3$) occurs when the temperature reaches 400°C, but with the use of the catalyst manganese dioxide (MnO_2) decomposition occurs at a temperature of only 200°C. Hydrogen peroxide decomposes with no heating at all as shown in the demonstration on this page.

There are many other examples of catalysts. One of the most common examples is the catalytic converter found in the exhaust systems of modern cars.

Making oxygen using electricity

Water contains 90% oxygen by mass. But the oxygen and hydrogen atoms are so tightly bound that it is difficult to imagine liquid water composed of elements that are normally found as gases.

To produce oxygen from water, a large amount of energy has to be applied in order to break the bonds between the oxygen and the hydrogen atoms. This is usually done with an electric current. In the demonstration on this page, water is split into its two elemental forms in an apparatus called Hoffman's voltameter. An industrial version of this electrical method is used for producing very pure oxygen.

Hoffman's voltameter

This equipment uses the principle of electrolysis; that is, by passing an electric current through a liquid, the compound can be forced to decompose (break up) into its building blocks.

It consists of three tubes. Each of the outer tubes has an electrode connected to an electrical supply.

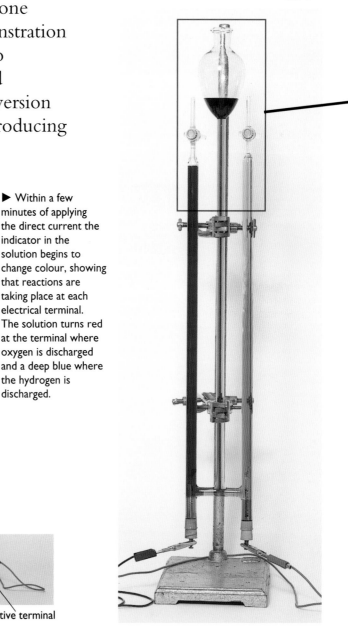

▼ Hoffman's voltameter consists of a direct current power supply and a set of three tubes connected together. The liquid in the tubes is water containing a chemical indicator to make it easier to see what processes are occurring and a dissolved substance (in this case sodium sulphate) that acts as an electrolyte and allows the current to flow. The change in colour of the indicator helps to identify the changes that take place.

The apparatus is filled with sodium sulphate solution containing an indicator (in this case Universal Indicator). Before the direct current is applied the whole solution is green.

The apparatus is held in place with a retort stand.

Direct current power supply.

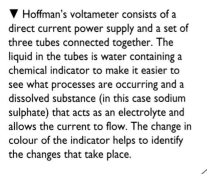

▶ Within a few minutes of applying the direct current the indicator in the solution begins to change colour, showing that reactions are taking place at each electrical terminal. The solution turns red at the terminal where oxygen is discharged and a deep blue where the hydrogen is discharged.

negative terminal   positive terminal

▼ This picture shows a detail of the Hoffman's voltameter, some time after the power has been applied. The indicator measures the amount of hydrogen ions in the water, that is, the amount of water that has been dissociated so that the water and hydrogen ions can move freely. The hydrogen ions recombine to hydrogen gas molecules in one tube (above the blue-stained water). The oxygen forms in the tube where the indicator has turned the water red.

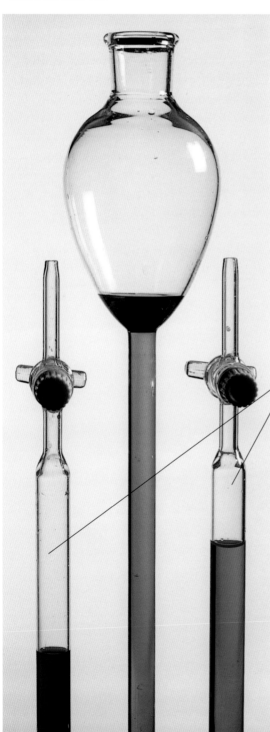

bond: chemical bonding is either a transfer or sharing of electrons by two or more atoms. There are a number of types of chemical bond, some very strong (such as covalent bonds), others weak (such as hydrogen bonds). Chemical bonds form because the linked molecule is more stable than the unlinked atoms from which it formed. For example, the hydrogen molecule (H_2) is more stable than single atoms of hydrogen, which is why hydrogen gas is always found as molecules of two hydrogen atoms.

dissociate: to break apart. In the case of acids it means to break up forming hydrogen ions. This is an example of ionisation. Strong acids dissociate completely. Weak acids are not completely ionised and a solution of a weak acid has a relatively low concentration of hydrogen ions.

electrolysis: an electrical–chemical process that uses an electric current to cause the break up of a compound and the movement of metal ions in a solution. The process happens in many natural situations (as for example in rusting) and is also commonly used in industry for purifying (refining) metals or for plating metal objects with a fine, even metal coating.

electrolyte: a solution that conducts electricity.

ion: an atom, or group of atoms, that has gained or lost one or more electrons and so developed an electrical charge.

Two volumes of hydrogen collect for every one of oxygen (the formula for water is H_2O).

EQUATION: Dissociation of water to yield oxygen

Water ⇨ hydrogen + oxygen

$$2H_2O(l) \quad ⇨ \quad 2H_2(g) \; + \; O_2(g)$$

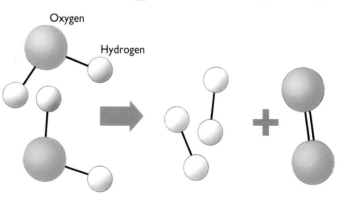

Oxygen

Hydrogen

Also...

Pure water, that is, water consisting only of water molecules with no impurities, is a poor conductor of electricity. However, the water we normally call pure water (meaning fit to drink) has many minerals dissolved in it. These minerals will dissociate easily, allowing ions to flow through the water and electrons to move through the wires of the external circuit.

Oxidising agents

Oxidising agents are commonplace around the home, in uses as diverse as cleaning wounds, sticking objects together and bleaching hair. Here are some examples of domestic oxidation reactions in action.

▼ The two components of this wood adhesive include a resin and an oxidising agent.

Adhesives

Many adhesives work by mixing two components together to produce a quick-setting compound. Often the reaction is between a resin (a dough-like material that can be moulded in the hands) and an oxidising agent, which is able to initiate the polymerising reaction and which is kept carefully in a tube. This is a very powerful chemical reaction and the oxidising agent is quite able to oxidise (corrode) skin, which accounts for the caution labels on the tube.

Many other adhesives set more slowly. In this case the oxygen from the air is used to oxidise the adhesive (a form of liquid plastic), often causing the adhesive units to gather together in a tangled mass of chains. This is the process of polymerisation, and it happens commonly with carbon-based adhesives, the kind that come in a tube and have a strong solvent smell (and caution remarks telling you not to inhale the fumes in a closed room).

Resin

Oxidising agent

Hair bleaches and detergents

Hair contains a number of colouring compounds (pigments). Hydrogen peroxide (which contains two hydrogen and two oxygen atoms) is an oxidising agent that can be used to lighten the colour of hair. A 6% solution of hydrogen peroxide is used for this purpose. The hydrogen peroxide reacts with the pigments in the hair, oxidising them to a colourless form.

Many detergents contain oxidising agents whose purpose is to turn coloured materials into colourless ones. In this way detergents are "stain" removers.

oxidation/reduction: a reaction in which oxygen is gained/lost. (Also… More generally oxidation or reduction involves the loss or gain of electrons.)

pigment: any solid material used to give a liquid a colour.

polymerisation: a chemical reaction in which large numbers of similar molecules arrange themselves into large molecules, usually long chains. This process usually happens when there is a suitable catalyst present. For example, ethene reacts to form polythene in the presence of certain catalysts.

resin: natural or synthetic polymers that can be moulded into solid objects or spun into thread.

◀▶ A sample of dark hair before and after treatment with hydrogen peroxide. The oxidation of the hair pigments has caused it to become much lighter.

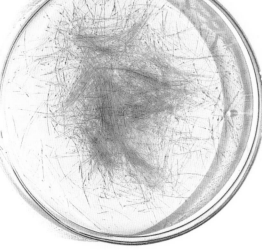

▶ Paint additives often contain an oxidising agent so that the paint will harden ("dry") without producing "runs". For example, driers, such as compounds of cobalt and manganese, are used in paints containing vegetable oils. This causes the pigment to absorb oxygen and polymerise, so hardening faster.

Cleaning wounds

Hydrogen peroxide (in a very dilute, 3% solution) decomposes in the presence of blood to form water and release oxygen gas. The oxygen reacts with any microorganism, killing it by oxidation and thereby disinfecting the wound.

Potassium permanganate (a violet-coloured potassium compound containing a large proportion of oxygen) works in much the same way.

Colours

Metal oxides are useful colouring agents and they have been used since ancient times as the pigment (coloured material) in paint.

An oxide of a particular metal may also have various colours, depending on how much oxygen is combined with the metal. For example, iron oxides can be red (when combined with much oxygen) or black (when combined with a smaller amount of oxygen).

The same rule applies to other metals, as you can see by the lead shown on this page.

▲ Lead oxides make a variety of colours. Massicot, a form of lead monoxide (PbO), is a yellow, crystalline compound. It is one of the most widely used and commercially important metallic compounds. The bright red powder is known as "red lead" (Pb_3O_4), whilst the darkest powder is lead dioxide (PbO_2).

Also:
Vanadium colours

Oxidation does not necessarily involve adding oxygen. It can also be defined as "loss of electrons". Elements are said to have "oxidation states", described by a number indicating the number of electrons lost. Each oxygen atom attached to an element corresponds to 2 lost electrons. Vanadium is a good example of a metal element that can be changed to several oxidation states, which have different colours as in the sequence of pictures shown below. Vanadium with the lowest oxidation state (2) is violet; with an oxidation state of 3 it is green; blue corresponds to an oxidation state of 4 and yellow to 5. The other colours are intermediate situations where two oxidation states are present (e.g. yellow plus blue appears green).

▼▶ The series of photographs across the bottom of these two pages shows the changing oxidation states of vanadium.

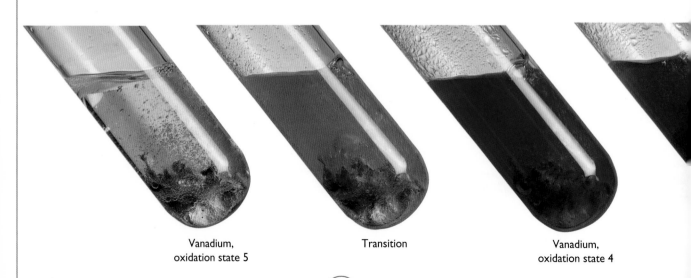

Vanadium, oxidation state 5

Transition

Vanadium, oxidation state 4

This black copper oxide is cupric oxide (CuO).

▼ Some oxides of copper and iron.

Most iron ore contains an intermediate amount of oxygen (Fe_3O_4) and is black.

This dark red iron oxide contains the most oxygen of any iron oxide. It is called ferric oxide (Fe_2O_3).

This is a red copper oxide, also known as cuprous oxide (Cu_2O).

Transition

Vanadium, oxidation state 3

Transition

Vanadium, oxidation state 2

Oxidation/reduction

Many of the reactions involving oxygen are reversible. That is, it is possible to add oxygen to a substance, a process called oxidation, and it is also possible to remove oxygen from a substance, a process called reduction.

Often the reversible process is called a redox (**red**uction-**ox**idation) reaction. Here we demonstrate the principle of redox reactions and where to find them.

Also...Lead-acid batteries

A lead-acid battery consists of alternating lead and lead oxide plates soaked in an electrolyte of sulphuric acid. When the battery is connected to a load (for example a vehicle starter motor), oxidation occurs at the lead anode and complementary reduction occurs at the lead oxide anode. Both changes release ions and so produce electricity. The chemical process is reversed when a generator is attached to the battery.

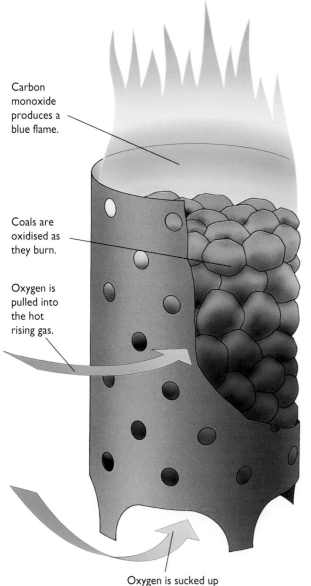

Carbon monoxide produces a blue flame.

Coals are oxidised as they burn.

Oxygen is pulled into the hot rising gas.

Oxygen is sucked up into the hot coals.

Oxidation and reduction in a fire

A fire in a brazier, a barbecue or a coal fire, shows the way that oxidation and reduction work.

Oxidation
The proof of the reduction process is in the blue flame. This is carbon monoxide burning. But the burning process is another oxidation process that finally produces carbon dioxide gas.

EQUATION: Carbon monoxide oxidised
Carbon monoxide + oxygen ⇨ carbon dioxide

$$2CO(g) \quad + \quad O_2(g) \quad ⇨ \quad 2CO_2(g)$$

Reduction
In this zone there is no more uncombined air and the carbon dioxide reacts with the carbon in the coals to produce carbon monoxide gas.

EQUATION: Carbon dioxide reduced
Carbon dioxide + carbon ⇨ carbon monoxide

$$CO_2(g) \quad + \quad C(s) \quad ⇨ \quad 2CO(g)$$

Oxidation
Carbon from the coals is oxidised to carbon dioxide gas. This reaction gives out heat (it is exothermic).

EQUATION: Coke is oxidised
Carbon + oxygen ⇨ carbon dioxide

$$C(s) \quad + \quad O_2(g) \quad ⇨ \quad CO_2(g)$$

Reduction/oxidation with copper

In the laboratory copper oxide (a black compound) can be reduced to copper (a reddish brown element) by passing carbon monoxide over the heated powder.

In this demonstration the gas carbon monoxide is passed through a special glass tube with a hole in the side. An excess of carbon monoxide is passed over the heated copper oxide and the unused carbon monoxide is burned off where it emerges from the hole.

The carbon monoxide reacts with the hot copper oxide and produces carbon dioxide gas. Carbon dioxide will not burn, so there is no flame at the outlet.

When all of the copper oxide has been converted to copper, carbon monoxide is flowing through the tube once more, and it ignites in the presence of the Bunsen flame.

The Bunsen flame is turned off, but carbon monoxide is still swept through the tube until the copper cools. If oxygen were to get in the tube while the copper was still hot, then the copper and the oxygen would react and reform the oxide. This would be a redox reaction.

exothermic reaction: a reaction that gives heat to the surroundings. Many oxidation reactions, for example, give out heat.

redox reaction: a reaction that involves reduction and oxidation.

Also...

Carbon monoxide is a very good reducing agent that will convert metal oxides to pure metal in a high temperature environment. Notice that it can be produced within a coke fire by drawing in a blast of air. This is the principle behind the iron-making blast furnace and the steel furnace, as you will see on later pages.

❶▼ Hot black copper oxide is being reduced by passing carbon monoxide over it. The oxygen from the copper reacts with the carbon monoxide gas to produce a noncombustible gas, carbon dioxide. As a result there is no flame at the outlet hole.

Oxidation of carbon monoxide and reduction of copper oxide

Copper oxide + carbon monoxide ➪ copper + carbon dioxide

$$CuO(s) \quad + \quad CO(g) \quad ➪ \quad Cu(s) \quad + \quad CO_2(g)$$

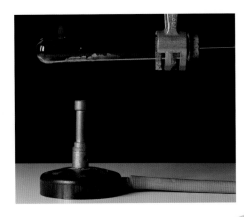

❷◄ The copper oxide is reduced to red copper. There is no more oxygen and so the carbon monoxide burns with blue flame.

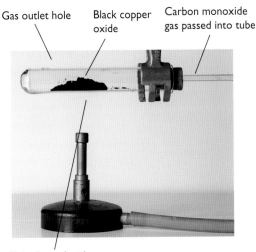

Gas outlet hole   Black copper oxide   Carbon monoxide gas passed into tube

Tube heated with a Bunsen burner flame

Notice that in this reaction the copper oxide has been reduced to copper and, at the same time, the carbon monoxide gas has been oxidised to carbon dioxide gas.

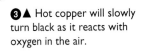

❸▲ Hot copper will slowly turn black as it reacts with oxygen in the air.

Oxygen in iron and steelmaking

Iron and steelmaking are perhaps the largest example of redox reactions in the world. In the iron-making process (which takes place in a blast furnace) the iron oxide has to be reduced; in the steelmaking process, the remaining carbon has to be oxidised so it will leave the iron as a gas and make steel. The amount of oxygen required is huge. It amounts to about 80 kilograms per person in most industrial countries.

The majority of the oxygen is used in steelmaking, where three-quarters of a tonne of oxygen is needed to make every tonne of steel.

How the blast furnace works

A blast furnace involves both reducing and oxidising environments, as the diagram below shows.

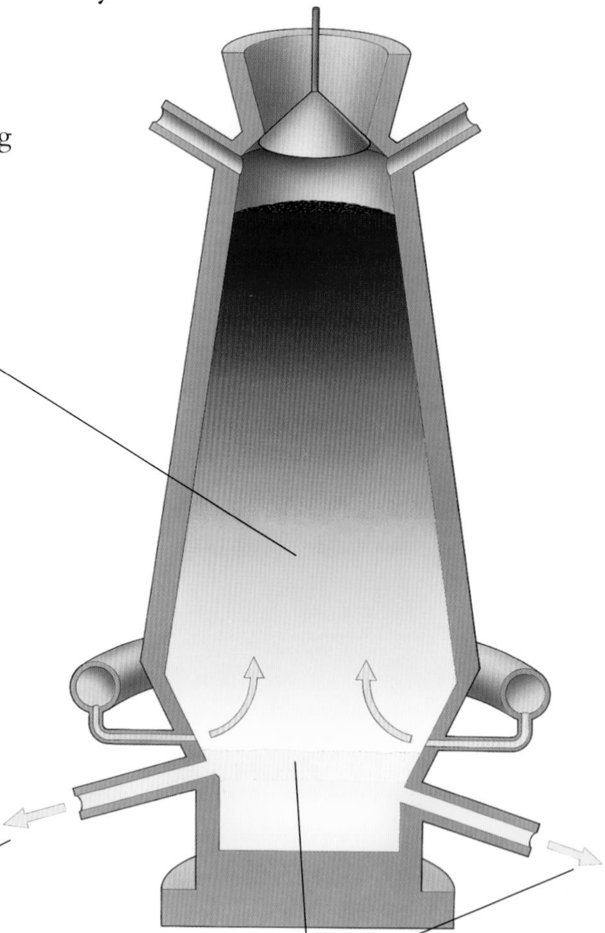

The carbon monoxide gas is produced at the bottom of the furnace. As air is blown in at the bottom of the furnace, so the coke, which is almost entirely carbon, is oxidised to produce carbon dioxide gas.

Carbon (solid) plus oxygen from the air (gas) react to give carbon dioxide (gas) and also release heat.

EQUATION: Oxidation of coke

Carbon + oxygen ⇨ carbon dioxide

$$C(s) \quad + \quad O_2(g) \quad ⇨ \quad CO_2(g)$$

EQUATION: Carbon dioxide reduced

Carbon dioxide + carbon ⇨ carbon monoxide

$$CO_2(g) \quad + \quad C(s) \quad ⇨ \quad 2CO(s)$$

The limestone in the charge decomposes to calcium oxide and gives off carbon dioxide gas. The calcium oxide reacts with the nonmetallic components of the ore. For example it reacts with silica of the rock to make calcium silicate.

The molten mixture of rock materials is known as slag. It is a light grey material, less dense than the iron, which is tapped off the furnace from holes above those used to tap the molten iron.

The iron oxide sinks down the furnace, where it reacts with carbon monoxide gas coming up from the bottom of the furnace. Carbon monoxide and iron oxide react to produce carbon dioxide gas and liquid iron.

EQUATION: Iron oxide reduced

Iron oxide + carbon monoxide ⇨ iron metal + carbon dioxide

$$Fe_2O_3(s) \quad + \quad 3CO(g) \quad ⇨ \quad 2Fe(l) \quad + \quad 3CO_2(g)$$

Steelmaking

The first steelmaking plant was known as the Bessemer Converter. A large metal container was lined with dolomite (calcium magnesium carbonate). When the container is heated the dolomite decomposed and combined with the impurities in the iron to make a slag that was drawn off. To help the process of oxidation further, air was blown through the container. The carbon in the iron oxidised to give carbon monoxide, a gas that then bubbled out of the steel.

The modern version of this process is called the basic oxygen process. Basic, because a base, limestone, is still used, and oxygen because pure oxygen, rather than air, is blown through the iron.

In the basic oxygen process the changes that occur are the same as in the Bessemer Converter, but a mixture of oxygen and powdered limestone is blown through the molten metal, thus making the chemical reactions happen faster and also more efficiently and evenly through the metal.

decompose: to break down a substance (for example by heat or with the aid of a catalyst) into simpler components. In such a chemical reaction only one substance is involved.

redox reaction: a reaction that involves reduction and oxidation.

slag: a mixture of substances that are waste products of a furnace. Most slags are composed mainly of silicates.

▲ This engraving shows a Bessemer Converter in use in the last century.

(i)

(ii)

Oxygen

(iii)

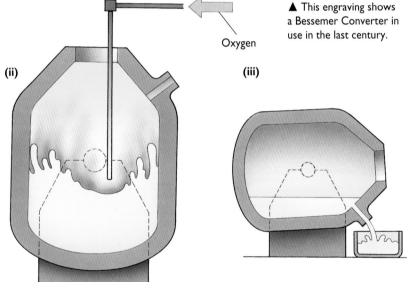

▲ The basic oxygen process. The furnace is charged with iron (i) and then oxygen is blown through the molten metal to form slag and to reduce the carbon content (ii). When the steel is at the desired chemical composition it is poured from the furnace (iii).

EQUATION: Iron is oxidised, releasing heat

Iron + oxygen ⇨ iron oxide

$$2Fe(s) + O_2(g) \Rightarrow 2FeO(s)$$

EQUATION: Excess carbon in the iron is removed

Iron oxide + carbon in the iron ⇨ steel + carbon monoxide

$$FeO(s) + C(s) \Rightarrow Fe(s) + CO(s)$$

Refining

From the chemist's point of view, ores can be refined in two ways. The simplest is to reduce or oxidise the ore (depending on the nature of the ore). This is what happens in an iron blast furnace (shown on the previous page) and in copper refining, shown on this page. But even then the reaction can only happen at high temperatures, showing that a large input of energy is needed.

Even the best of chemical reactions cannot completely remove all of the impurities in a metal, so ores refined in a furnace do not produce pure metals. Indeed, some metals are so reactive that they cannot be separated from their ores by chemical reactions at all. Aluminium and zinc are important examples. This is why many metals are refined to their final stage of purity by electrical means (a process called electrolysis).

The reactivity of metals

Metals can be arranged in a list, called a reactivity series, with the most reactive at the top and the least reactive at the bottom. Those with low reactivity are the easiest to refine; those with high reactivity are far more difficult. Thus aluminium, which is a reactive metal, can only be separated from its ore, bauxite (aluminium oxide), by the use of electrical energy, a process called electrolysis. On the other hand, iron and copper oxides are much easier to refine (at least until the highest purity is required) because copper and iron are less reactive metals. Here a high temperature is sufficient to ensure that the ore reacts.

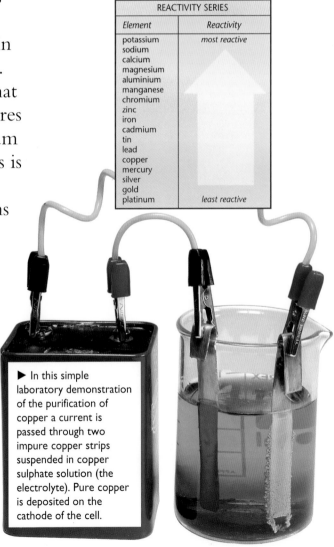

REACTIVITY SERIES	
Element	Reactivity
potassium	most reactive
sodium	
calcium	
magnesium	
aluminium	
manganese	
chromium	
zinc	
iron	
cadmium	
tin	
lead	
copper	
mercury	
silver	
gold	
platinum	least reactive

▼▶ The last stage of refining copper is done by electrolysis. The impure copper from the furnace is made into one electrode of an electrolysis cell. The other electrode is made from a thin sheet of pure copper. The copper is refined by placing the two electrodes in a copper sulphate bath and passing a current between them. The impure copper on the anode corrodes and pure copper collects on the cathode, When the cathode has acquired sufficient copper, it will be melted down and made into wires and sheet metal.

The giant industrial electrolysis and the laboratory equivalent are shown here.

▶ In this simple laboratory demonstration of the purification of copper a current is passed through two impure copper strips suspended in copper sulphate solution (the electrolyte). Pure copper is deposited on the cathode of the cell.

Refining copper

Copper occurs as both oxide ores and sulphide ores. The easiest to refine are the oxides. In an industrial furnace the carbon monoxide gas needed to reduce the ore (remove the oxygen from the metal) is produced by heating coke, a source of carbon, and feeding in a jet of air. The carbon reacts with oxygen from the air to form carbon monoxide, which then reduces the copper oxide to copper. The copper can then be tapped from the base of the furnace. This is similar to the iron blast furnace shown on pages 26–27.

Copper ores containing sulphur are very much more difficult to deal with, yet they make up half of all copper deposits. These ores have to be *oxidised*, as shown by the equation on this page. When they are burnt, huge quantities of polluting sulphur dioxide gas are produced. This gas is one of the major causes of acid rain, and modern factories recover as much as they can in the production of sulphuric acid.

electrolysis: an electrical–chemical process that uses an electric current to cause the break up of a compound and the movement of metal ions in a solution. The process happens in many natural situations (as for example in rusting) and is also commonly used in industry for purifying (refining) metals or for plating metal objects with a fine, even metal coating.

sulphide: a sulphur compound that contains no oxygen.

▼ A furnace of copper being unloaded after smelting.

EQUATION: Oxidation of copper sulphide ore in a smelter to produce copper

Copper sulphide + limited air supply ⇨ copper + sulphur dioxide

$$2CuS(s) \quad + \quad 2O_2(g) \quad \Rightarrow \quad 2Cu(l) \quad + \quad 2SO_2(g)$$

Corrosion

One of the most important properties of most oxides is that they are insoluble. Thus when a highly reactive metal, such as aluminium, is exposed to the air, it very quickly reacts with the oxygen in the air and develops a thin oxide coating.

The oxide coating is almost invisible and it is gas and watertight. This is what prevents further reaction with air (corrosion). It also explains why there is an apparent contradiction between the reactivity of a metal like aluminium, and its apparent immunity from corrosion.

One of the important consequences of such rapid oxidation is that many goods can be manufactured from a wide variety of metals without risk of them disintegrating quickly. Objects from toys to vehicles, houses to skyscrapers rely on metals developing a protective oxide coating.

Unfortunately, one of the most widely used metals, iron, does not develop a watertight oxide coating. This is the reason iron corrodes, a process known as rusting. As a result, while aluminium can be used for external cladding without treating it, steel must be covered with a coating (for example paint) to protect it from the effects of water.

▶ An iron nail placed in a jar of water quickly rusts.

EQUATION: The rusting of iron

Iron + water + oxygen ⇨ ferric oxide + water

$$4Fe(s) + 6H_2O(l) + 3O_2(g) \Rightarrow 2Fe_2O_3(s) + 6H_2O(l)$$
$$\text{Rust}$$

How iron becomes pitted

If a wetted iron surface is exposed either by being uncoated or because the paint on the surface has been chipped, oxygen atoms are able to enter the water through its surface skin. In this water one of the world's tiniest batteries forms. The water is oxygen-rich and the iron forms an electrode, one terminal of a battery. The oxygen-poor region, in the scratch and farther from the air, forms the other electrode. The water forms the electrolyte. A minute electric current now flows, and iron is carried in solution to the oxygen-rich water, where it is oxidised and deposited.

Thus, iron is removed from one part of the metal and deposited as an oxide or hydroxide nearby. This explains why rusty material is often both pitted and lumpy.

corrosion: the *slow* decay of a substance resulting from contact with gases and liquids in the environment. The term is often applied to metals. Rust is the corrosion of iron.

electrode: a conductor that forms one terminal of a cell.

electrolyte: a solution that conducts electricity.

▼ These rusting chains show two forms of rust. The light brown rust patches are recently formed iron hydroxide or $Fe(OH)_3$. In contrast, the darker brown rust patches are the final solid state of ferric oxide (Fe_2O_3).

Preventing corrosion

Corrosion is a major problem for anybody using iron products. One of the simplest ways to protect iron is to seal it from damp air. The most common sealants are paints. To be useful paints need to have "skins" that are gas and watertight.

Once iron has begun to corrode, covering it with paint will not stop the rusting. Anyone who has painted a rusty object will know that the paint soon bubbles up and the rust shows through. In fact the painting has *increased* the rate of rusting by making sure that the rust (and the moisture it contains) remains in contact with the iron. The only way to cure rust is to treat it chemically, so that the rust is converted to an unreactive substance.

Rusting is actually a process of electrolysis, as shown on page 31. Rusting can also, therefore, be prevented by operating electrolysis in reverse, a process called cathodic protection.

Corrosion-proof coating

Steel-hulled ships and motor vehicles are very prone to corrosion because they are continuously exposed to wet or damp conditions. The traditional way to protect them was to apply coats of paint, but despite this it was common for rust to bubble up under the paint because the paint was not well bonded to the metal.

In recent years better coating materials have been applied to the steel used in vehicles to prevent oxygen reaching the steel. Up to six coatings are now applied, beginning with a coating of iron phosphate and including several layers of plastic seal. These special coatings are the reason many vehicles now have anti-rust warranties lasting for several years.

Another way to coat the steel is to use a coating of a metal that is not prone to corrosion because the oxide that forms on its surface is gastight. Chrome, tin and zinc are commonly used for this purpose. Tin was traditionally used to protect food cans as "tin-plate", while zinc was mainly used for protecting steel exposed to the weather as "galvanised steel".

▶ Rust-inhibitors are compounds such as phosphoric acid. The reaction between the acid and iron oxide converts the iron oxide layer to iron phosphate, a glassy, insoluble material that is gastight when it dries out.

EQUATION: Phosphoric acid rust-inhibitor

Phosphoric acid + iron oxide ⇨ iron phosphate + water

$$2H_3PO_4(aq) \quad + \quad Fe_2O_3(s) \quad ⇨ \quad 2FePO_4(s) \quad + \quad 3H_2O(aq)$$

Galvanising

Galvanising is the process of coating iron with a surface of zinc metal. For some applications the metal is dipped in a bath of molten zinc; for others the zinc is applied in an electrolytic bath.

Zinc protects iron because it forms a gastight oxide. If the zinc coating is scratched, it still protects the iron because zinc is more reactive than iron.

▲ Galvanised wire used in a fence.

Cathodic protection

Cathodic protection is a form of corrosion-proofing. It relies on the fact that when two metals are placed in a liquid, they behave like a battery and electricity flows. During this process metal is lost from the negative electrode, called an anode, and it corrodes, or oxidises; at the positive electrode (called the cathode), reduction occurs and corrosion is impossible. By choosing metals carefully, a chemist can decide which will be the corroding anode and which the protected cathode.

A block of magnesium, zinc or other suitable metal is fastened on to the hull of a steel ship, or buried in the ground with the steel pipe and connected to it by conducting cables. Because magnesium and zinc are more reactive than iron, the blocks will act as anodes and corrode instead of the ship or pipe. Of course, they have to be replaced from time to time as part of routine maintenance.

anode: the negative terminal of a battery or the positive electrode of an electrolysis cell.

cathode: the positive terminal of a battery or the negative electrode of an electrolysis cell.

cathodic protection: the technique of making the object that is to be protected from corrosion into the cathode of a cell. For example, a material, such as steel, is protected by coupling it with a more reactive metal, such as magnesium. Steel forms the cathode and magnesium the anode. Zinc protects steel in the same way.

electrolysis: an electrical–chemical process that uses an electric current to cause the break up of a compound and the movement of metal ions in a solution. The process happens in many natural situations (as for example in rusting) and is also commonly used in industry for purifying (refining) metals or for plating metal objects with a fine, even metal coating.

electroplating: depositing a thin layer of a metal onto the surface of another substance using electrolysis.

reactivity: the tendency of a substance to react with other substances. The term is most widely used in comparing the reactivity of metals. Metals are arranged in a reactivity series.

Tin-plating

Steel sheet is dipped in tin to make tin-plate. Tin is an unreactive metal, which resists chemical reactions that would result in corrosion. However, because iron is more reactive than tin, if the tin-plate is scratched, the iron will corrode very quickly.

▶ Tin-plated can.

Combustion

Oxygen itself is noncombustible, even though one of the most striking characteristics of oxygen is how it reacts much more quickly at high temperatures compared with normal environmental temperatures. The chemical reaction is called oxidation, but the rapid effect is known as combustion.

A clear example of this is wood. At room temperature we can use wood for furniture, flooring and as roofs for our houses. At these low temperatures wood hardly reacts at all with oxygen, making it possible for wooden structures to last for many hundreds of years.

But if we increase the temperature of the wood, the rate of reaction with air increases dramatically. Thus we can use wood as a fuel to burn. At these higher temperatures, the chemical reaction that occurs gives out a large amount of heat.

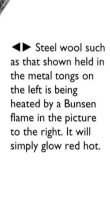

◄► Steel wool such as that shown held in the metal tongs on the left is being heated by a Bunsen flame in the picture to the right. It will simply glow red hot.

► Heated steel wool burning in a gas jar of oxygen. Notice the steel is burning (there is no other source of flame) and steel droplets are falling to the bottom of the gas jar. The water in the gas jar is there to douse the droplets of iron oxide as they fall.

EQUATION: Burning steel wool

Iron + oxygen ⇨ iron oxide

$4Fe(s) + 3O_2(g) \Rightarrow 2Fe_2O_3(s)$

Fire extinguishers and flame retardants

Because combustion is caused by the rapid oxidation of a material, most ways of treating a fire involve two stages: to reduce the temperature, thus slowing down the rate of reaction, and to deny the fuel a further supply of oxygen.

A simple fire blanket thrown over a fire helps to starve a fire of the oxygen needed for continued burning. Water acts to reduce the temperature and also to cover the fuel so that very little air can reach it. Carbon dioxide gas cannot burn; using carbon dioxide fire extinguishers covers the fire and thus keeps the oxygen out. Powder extinguishers are made of a noncombustible powder that simply blankets the burning material and stops oxygen getting to the fuel.

A flame-retardant is a chemical used on clothing and furniture fabrics that prevents a material from burning easily. The most common chemicals are based on phosphoric or sulphuric acid. When the chemical begins to burn, the compound decomposes, leaving an acid behind that can react with the fabric. The effect is to reduce the organic material of the fabric to carbon. The carbonised (charred) surface produces little fuel with which the oxygen in the air can react. The same chemical reaction releases carbon dioxide, again preventing oxygen from getting to the material. Borax is another material sometimes used as a flame-retardant coating. Borax melts at fairly low temperatures, smothering the fabric and keeping oxygen from the inflammable fabric.

The only problem with these materials is that they release toxic gases as they react, thereby making it necessary for firefighters to use breathing apparatus.

EQUATION: Fire extinguishing

Aluminium sulphate + sodium carbonate + water ⇨ aluminium hydroxide + carbon dioxide + sodium sulphate

$$Al_2(SO_4)_3(s) + 3Na_2CO_3(aq) + 3H_2O \Rightarrow 2Al(OH)_3(s) + 3CO_2(g) + 3Na_2SO_4(aq)$$

Spontaneous combustion

Spontaneous combustion is the effect of a material beginning to oxidise *very quickly*. An oxidising agent, for example, readily reacts with another substance, giving up its oxygen. The reaction that transfers oxygen often causes the release of considerable amounts of heat (it is an exothermic reaction).

Demonstration of delayed oxidation

Potassium permanganate (an oxidising agent) and ethylene glycol (antifreeze) show the delayed effects of some chemical reactions. On their own these chemicals are harmless and will not catch fire. However, when mixed together they can be extremely dangerous.

When first mixed nothing appears to happen, because the chemicals need time to react. However, all the time the temperature is rising. Chemists call this waiting time an induction period. After a minute or two the chemicals suddenly reach a temperature where the reaction happens very fast indeed, and the mixture appears to burst into flame spontaneously.

❶▼ A few drops of ethylene glycol are added to crystals of potassium permanganate.

❷▲ A few minutes later the reaction speeds up and flames begin to shoot into the air.

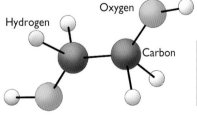

Hydrogen Oxygen Carbon

▲ A molecule of ethylene glycol

EQUATION: Ethylene glycol and potassium permanganate

Ethylene glycol + oxidising agent ⇨ carbon dioxide+ water

$C_2H_6O_2(l) + 5O$ (from oxidising agent) ⇨ $2CO_2(g) + 3H_2O(l)$

Spontaneous combustion

Spontaneous combustion can be due to combining an oxidising agent with another substance as shown on these pages. Matches are a chemical reaction between the chemicals of the match head and the box side. The result is rapid oxidation and the generation of enough heat to catch the match wood alight.

Spontaneous combustion can also occur when, for example, a strong light is focused by a piece of glass. This will cause the temperature of a sheet of paper to rise enough for rapid oxidation (burning) to occur. It shows clearly that a flame is not needed to set something alight.

Some people find out that microwaves can also cause combustion. If the timer on a microwave oven is set too long, the food may get hot enough to catch fire.

Spontaneous combustion can occur in any place where there is poor circulation of air around a fuel. For example, dried grass in a haystack will oxidise, releasing heat. If there is enough air to allow continued oxidation, but not enough to carry away the heat, then a temperature may be reached when combustion occurs. Piles of oily rags and coal have all been known to burst into flame spontaneously.

Spontaneous combustion is also possible with any dusty fuel, because the greater surface area of the dust means that it can oxidise more easily. This is why coal dust explodes in mines and why even flour dust in a flour mill sometimes catches fire.

exothermic reaction: a reaction that gives heat to the surroundings. Many oxidation reactions, for example, give out heat.

induction period: the time that sometimes elapses between the start of a chemical reaction and when that reaction actually becomes obvious. For example, a fabric might begin to smoulder imperceptibly if heated and then, sometime later, burst into flames without any apparent cause. Chemical reactions can be extremely dangerous if people are unaware of the induction period involved.

spontaneous combustion: the effect of a very reactive material beginning to oxidise very quickly and bursting into flame.

❸ ▼ The potassium permanganate and ethylene glycol have produced enough heat to make a miniature "volcano" in the laboratory.

Combustion of fuels

When a material burns it undergoes a chemical reaction (combustion) that produces new products and gives out heat. Some materials give out more heat than others. Those that give out the most heat for the smallest volume are called fuels.

The products of fuel combustion are usually carbon dioxide and water vapour, together with any compounds of nitrogen and sulphur that might have been in the fuel as impurities. Solid particles of carbon and carbon monoxide gas may also be produced if the combustion was not complete.

The volume and mass of the products are greater than the volume and mass of the fuel. This is because oxygen is taken from the air. For example, when one hundred grams of propane fuel is burned it releases three hundred grams of carbon dioxide. This is because the product contains two hundred grams of invisible oxygen taken from the air.

The temperatures at which combustion occurs vary widely. For example, kerosene (aviation fuel) burns at about 300°C, wood and petrol at 350°C, coal at 400°C, hydrogen (rocket fuel) at about 600°C and aluminium at 2000°C.

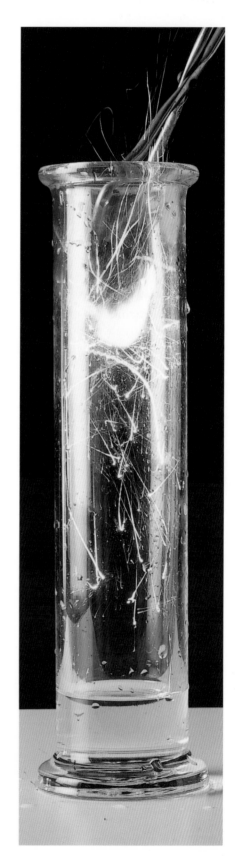

◀ ▶ These pictures show that charcoal will simply glow in air, even when heated strongly by a Bunsen burner. However, when the hot charcoal is placed in a gas jar containing oxygen, the charcoal combusts and a ball of white-hot carbon is produced. Carbon dioxide is given off until the oxygen is used up.

Propulsion in space

Fuels burn because they have a supply of oxygen. In space there is no oxygen, so this must be provided. All spacecraft – rockets, space shuttles, etc. – must have their own on-board supply of oxygen. This can be supplied as liquid oxygen or as an oxidising chemical.

The strongest of all common acids is called perchloric acid. If even a small amount of perchloric acid gets into contact with any organic matter, the reaction is so swift that it can cause an explosion. This oxidation potential can be used to great advantage under controlled conditions. In the booster rockets of the space shuttle, ammonium perchlorate and aluminium powder are packed together. The fuel (aluminium powder) thus has its own oxygen supply as ammonium perchlorate

As it burns, the mixture throws out clouds of white aluminium oxide powder, hydrochloric acid gas, nitrogen oxide gas and water vapour. It is the rapid expansion of these gases that provides the thrust for the space shuttle and the white trail that you can see.

◀ A space shuttle taking off. The brown tank underneath the space shuttle contains liquid oxygen and hydrogen fuel.

combustion: the special case of oxidisation of a substance where a considerable amount of heat and usually light are given out. Combustion is often referred to as "burning".

fuel: a concentrated form of chemical energy. The main sources of fuels (called fossil fuels because they were formed by geological processes) are coal, crude oil and natural gas. Products include methane, propane and gasoline. The fuel for stars and space vehicles is hydrogen.

welding: fusing two pieces of metal together using heat.

◀ Oxyacetylene welding in progress.

Oxyacetylene cutting and welding

Liquid oxygen is used in bottled form for oxyacetylene cutting and welding.

In this process, the acetylene fuel (kept liquefied in a cylinder) is fed into a nozzle, where it combines with oxygen supplied from a separate cylinder. When this mixture is ignited the reaction releases a large amount of heat, while producing carbon dioxide gas and water vapour.

Oxyacetylene torches can reach temperatures of over 3300°C. This flame is so hot it can melt metal, which is why it is used in welding.

By changing the nature of the oxygen and acetylene head, the torch can be made to cut metal. In this case a jet of oxygen is introduced into the centre of the acetylene flame. This oxidizes the white-hot metal, leaving a narrow, clean-edged cut.

Internal combustion

One of the most widely used fuels is petrol or gasoline, a carbon-based liquid that evaporates easily to release fumes. If the temperature of the gasoline is raised to over 350°C, as for example, by bringing a lighted match near to it, then the fumes (vapour) evaporating from the liquid will react with the oxygen in the surrounding air, creating carbon dioxide and water vapour and at the same time giving out a lot of heat.

   If the fuel is formed into a mist of tiny droplets (it is "atomised"), such as happens to fuel being sucked into the cylinder of an engine, the fuel will burn much faster (see page 38). When the mist is ignited with a spark from a spark plug, combustion occurs and the liquid droplets (where the molecules are tightly packed together) are transformed into a gas where the molecules are widely spaced. The rapid expansion of the gas (a controlled explosion) forces the piston down the cylinder, driving the vehicle, and also forces the spent gases out through the exhaust system. This process is called internal combustion.

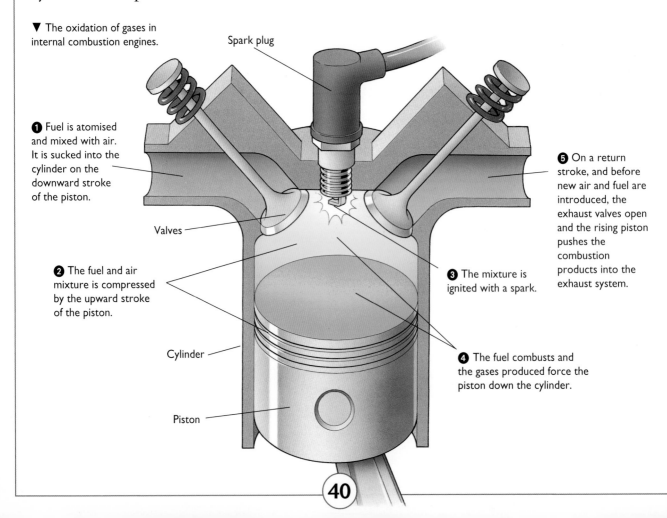

▼ The oxidation of gases in internal combustion engines.

Spark plug

❶ Fuel is atomised and mixed with air. It is sucked into the cylinder on the downward stroke of the piston.

❷ The fuel and air mixture is compressed by the upward stroke of the piston.

Valves

Cylinder

Piston

❸ The mixture is ignited with a spark.

❹ The fuel combusts and the gases produced force the piston down the cylinder.

❺ On a return stroke, and before new air and fuel are introduced, the exhaust valves open and the rising piston pushes the combustion products into the exhaust system.

Environmental pollution

The combustion that occurs inside a motor engine is not very efficient. If it were completely efficient then all the fuel would be changed into carbon dioxide and water and two harmless gases would escape from the exhaust pipes of vehicles. However, because the fuel is not completely burned up (oxidised) carbon monoxide is formed and many hydrocarbon particles are ejected from the partly burnt fuel.

When cars idle they are at their least efficient, so idling cars produce more carbon monoxide than cars travelling quickly. This is why carbon monoxide pollution is higher in cities (where many cars are moving slowly) than in the countryside. (See also the discussion of smog on the next page for nitrogen dioxide pollution.)

Catalytic converter

A catalytic converter is a device that looks like an additional silencer set into the exhaust systems of modern petrol vehicles.

The catalytic converter is actually a chemical device, designed to cause a range of reactions with the exhaust gases of the engine (carbon monoxide and nitrogen oxides) and hydrocarbon fragments and convert these potentially harmful pollutants into the harmless, naturally occurring products: nitrogen, carbon dioxide and water.

The catalyst is called an oxidation/reduction catalyst because it causes the oxidation of carbon monoxide and hydrocarbon fragments to carbon dioxide and water, using the oxygen from the nitrogen oxides, thus converting them to inert nitrogen gas (reduction).

The catalyst is a combination of about three grams of the rare elements platinum and palladium, which explains why the catalytic converters are expensive. The catalyst is fixed to the surface of some supporting structure, such as a porous ceramic material (whose total surface area is equivalent to two soccer pitches). This provides the most efficient exposure of the catalyst to the gases.

EQUATION: Oxidation of carbon monoxide

Carbon monoxide + oxygen ⇨ carbon dioxide

$2CO(g)$ + $2O$ (from NO_x) ⇨ $2CO_2(g)$

EQUATION: Reduction of nitric oxide

Nitric oxide ⇨ nitrogen + oxygen

$2NO(g)$ ⇨ $N_2(g)$ + $O_2(g)$

atomised: broken up into a very fine mist. The term is used in connection with sprays and engine fuel systems.

catalyst: a substance that speeds up a chemical reaction but itself remains unaltered at the end of the reaction.

ceramic: a material based on clay minerals, which has been heated so that it has chemically hardened.

Combustion of powders and mists

Chemical reactions work faster when the surface area of contact between the two reagents is high. Chemical reactions work very quickly when two liquids are mixed because the molecules of the compounds are in complete contact. The fine mist of fuel in a vehicle engine reacts quickly because the surface area of the fuel is large (see page 40).

A large block of coal burns slowly because there is little surface area with which the oxygen can react. By contrast, a powder of coal dust will combust so quickly that it will create an explosion. Other organic materials will behave in the same way, for example flour dust in a flour mill (flour is a fuel because it is made of a carbon compound). This is the reason coal mine and grain store managers have to be so careful not to allow the build-up of dust or the use of flames or sparks.

Also...

A car has an engine that creates chemical reactions thousands of times each minute. Each of the reactions is a form of combustion, mixing a fuel and oxygen at a high temperature inside a closed cylinder.

Because the reaction happens inside the cylinders of motor vehicles, the engines are called internal combustion engines.

The mixture of oxygen and gasoline is brought to a high temperature partly by squeezing it inside the cylinder and partly by igniting a spark in the top of the cylinder. But diesel engines do not use spark plugs at all. They simply compress the mixture until it gets so hot that the chemical reaction occurs.

Because it is atomised, most of the fuel is able to undergo combustion at the same time.

Smog: oxidation of exhaust fumes

Oxidation of exhaust gases is one of the main sources of pollution in the world. For example, a brown haze often hangs over some of the world's major cities, especially those, like Los Angeles, Mexico City and Bangkok, that have heavy traffic, bright sunshine and calm air. This haze is called smog (or more properly, photochemical smog).

Smog is a result of chemical reactions that take place in the air, using the energy of sunlight. Smog begins when gases are created in the cylinders of vehicle engines. Here oxygen and nitrogen gas combine as the fuel burns to form nitric oxide (NO, a colourless gas). Together with a cocktail of other gases, nitric oxide is pumped out of the tailpipes of vehicles.

When the gas reaches the air, more oxygen is available from the atmosphere. The nitric oxide combines with oxygen to produce nitrogen dioxide (NO_2, a brown gas). This is the gas that will eventually contribute to acid rain in wet environments. However, in dry sunny places the intense sunlight causes the nitrogen dioxide to decompose back into nitric oxide and releases oxygen atoms (O).

The formation of nitrogen dioxide is in part responsible for the brown colour of the air. But the released oxygen atoms are very reactive and quickly combine with oxygen (O_2) molecules to form ozone (O_3).

Ozone (see page 12) is highly toxic and even at tiny concentrations causes irritation to the eyes, especially in congested city streets on hot sunny days.

Demonstration of the smog effect

If a gas jar of nitric oxide is put together with a gas jar of air, and the separating cover slip pulled away, the nitric oxide will mix with the air and oxidise to brown nitrogen dioxide.

Because gas molecules are in constant motion, the contents of both gas jars become uniformly brown.

▼ Photochemical smog over Los Angeles, California, USA.

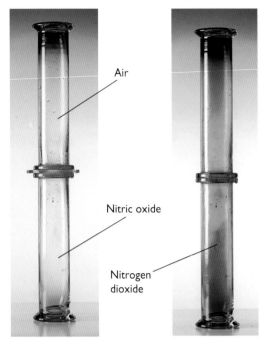

▼ Laboratory demonstration showing the mixing of nitric oxide and air to produce nitrogen dioxide.

Air

Nitric oxide

Nitrogen dioxide

decompose: to break down a substance (for example by heat or with the aid of a catalyst) into simpler components. In such a chemical reaction only one substance is involved.

ozone: a form of oxygen whose molecules contain three atoms of oxygen. Ozone is regarded as a beneficial gas when high in the atmosphere because it blocks ultraviolet rays. It is a harmful gas when breathed in, so low level ozone, which is produced as part of city smog, is regarded as a form of pollution. The ozone layer is the uppermost part of the stratosphere.

toxic: poisonous enough to cause death.

Exhaust fumes

Nitrogen dioxide decomposes to release nitric oxide and oxygen atoms.

EQUATION: Result of nitrogen dioxide decomposing to nitric oxide and oxygen (under the influence of strong light energy)

Nitrogen dioxide ⇨ nitric oxide + oxygen atoms

$$NO_2(g) \quad ⇨ \quad NO(g) \quad + \quad O(g)$$

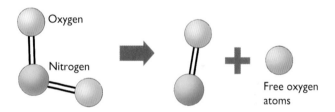

Oxygen

Nitrogen

Free oxygen atoms

EQUATION: O plus O_2 (oxygen from the air) react to make the irritant in smog, O_3

Oxygen atoms + oxygen molecules ⇨ ozone molecules

$$O(g) \quad + \quad O_2(g) \quad ⇨ \quad O_3(g)$$

Free oxygen atoms

Many of the free oxygen atoms combine with molecules of oxygen gas (O_2) to produce the toxic gas, ozone (O_3).

Key facts about...

Oxygen

A colourless gas, chemical symbol O

Found in two gaseous forms O_2 and O_3 (ozone)

Most plentiful element at the surface of the Earth

Important for burning fuels

Essential for breathing

Combines with hydrogen to make water

Has no taste

Has no smell

Forms the minerals that make rocks

Essential for converting food into energy

Part of the tissues of all living organisms

Makes up 21% of the atmosphere

Atomic number 8, atomic weight about 16

SHELL DIAGRAMS

The shell diagram on this page represents an atom of the element oxygen. The total number of electrons is shown in the relevant orbitals, or shells, around the central nucleus.

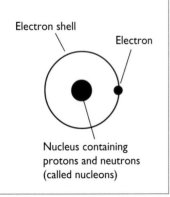

Electron shell

Electron

Nucleus containing protons and neutrons (called nucleons)

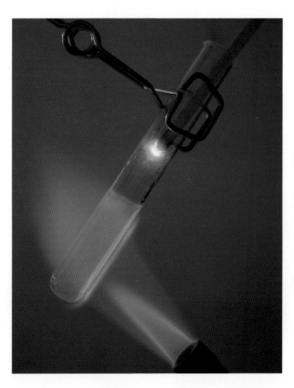

◀▶ The standard laboratory test for the presence oxygen is that it will relight a glowing splint. In this demonstration potassium nitrate is being melted in a test tube. The reaction that takes place produces a greenish yellow liquid of potassium nitrite and gives off oxygen gas that can be seen as bubbles in the solution.

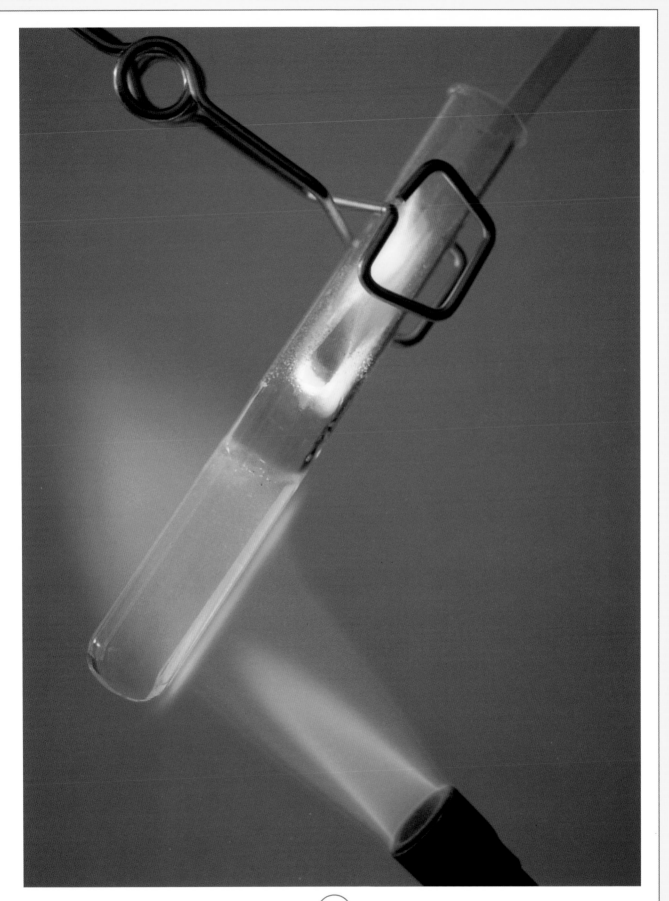

The Periodic Table

The Periodic Table sets out the relationships among the elements of the Universe. According to the Periodic Table, certain elements fall into groups. The pattern of these groups has, in the past, allowed scientists to predict elements that had not at that time been discovered. It can still be used today to predict the properties of unfamiliar elements.

The Periodic Table was first described by a Russian teacher, Dmitry Ivanovich Mendeleev, between 1869 and 1870. He was interested in writing a chemistry textbook, and wanted to show his students that there were certain patterns in the elements that had been discovered. So he set out the elements (of which there were 57 at the time) according to their known properties. On the assumption that there was pattern to the elements, he left blank spaces where elements seemed to be missing. Using this first version of the Periodic Table, he was able to predict in detail the chemical and physical properties of elements that had not yet been discovered. Other scientists began to look for the missing elements, and they soon found them.

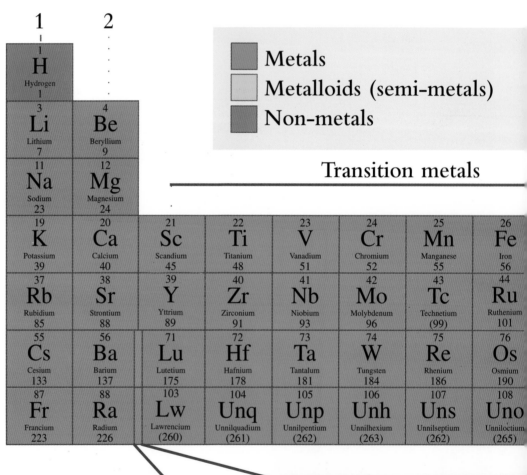

GROUP

Metals
Metalloids (semi-metals)
Non-metals

Transition metals

Lanthanide metals

Actinoid metals

1	2						
1 H Hydrogen 1							
3 Li Lithium 7	4 Be Beryllium 9						
11 Na Sodium 23	12 Mg Magnesium 24						
19 K Potassium 39	20 Ca Calcium 40	21 Sc Scandium 45	22 Ti Titanium 48	23 V Vanadium 51	24 Cr Chromium 52	25 Mn Manganese 55	26 Fe Iron 56
37 Rb Rubidium 85	38 Sr Strontium 88	39 Y Yttrium 89	40 Zr Zirconium 91	41 Nb Niobium 93	42 Mo Molybdenum 96	43 Tc Technetium (99)	44 Ru Ruthenium 101
55 Cs Cesium 133	56 Ba Barium 137	71 Lu Lutetium 175	72 Hf Hafnium 178	73 Ta Tantalum 181	74 W Tungsten 184	75 Re Rhenium 186	76 Os Osmium 190
87 Fr Francium 223	88 Ra Radium 226	103 Lw Lawrencium (260)	104 Unq Unnilquadium (261)	105 Unp Unnilpentium (262)	106 Unh Unnilhexium (263)	107 Uns Unnilseptium (262)	108 Uno Unniloctium (265)

57 La Lanthanum 139	58 Ce Cerium 140	59 Pr Praseodymium 141	60 Nd Neodymium 144
89 Ac Actinium (227)	90 Th Thorium 232	91 Pa Protactinium 231	92 U Uranium 238

Hydrogen did not seem to fit into the table, so he placed it in a box on its own. Otherwise the elements were all placed horizontally. When an element was reached with properties similar to the first one in the top row, a second row was started. By following this rule, similarities among the elements can be found by reading up and down. By reading across the rows, the elements progressively increase their atomic number. This number indicates the number of positively charged particles (protons) in the nucleus of each atom. This is also the number of negatively charged particles (electrons) in the atom.

The chemical properties of an element depend on the number of electrons in the outermost shell.

Atoms can form compounds by sharing electrons in their outermost shells. This explains why atoms with a full set of electrons (like helium, an inert gas) are unreactive, whereas atoms with an incomplete electron shell (such as chlorine) are very reactive. Elements can also combine by the complete transfer of electrons from metals to non-metals and the compounds formed contain ions.

Radioactive elements lose particles from their nucleus and electrons from their surrounding shells. As a result their atomic number changes and they become new elements.

Key to diagram:

- Atomic (proton) number — 13
- Symbol — Al
- Name — Aluminium
- Approximate relative atomic mass (Approximate atomic weight) — 27

3	4	5	6	7	0
					2 **He** Helium 4
5 **B** Boron 11	6 **C** Carbon 12	7 **N** Nitrogen 14	8 **O** Oxygen 16	9 **F** Fluorine 19	10 **Ne** Neon 20
13 **Al** Aluminium 27	14 **Si** Silicon 28	15 **P** Phosphorus 31	16 **S** Sulphur 32	17 **Cl** Chlorine 35	18 **Ar** Argon 40

27 **Co** Cobalt 59	28 **Ni** Nickel 59	29 **Cu** Copper 64	30 **Zn** Zinc 65	31 **Ga** Gallium 70	32 **Ge** Germanium 73	33 **As** Arsenic 75	34 **Se** Selenium 79	35 **Br** Bromine 80	36 **Kr** Krypton 84
45 **Rh** Rhodium 103	46 **Pd** Palladium 106	47 **Ag** Silver 108	48 **Cd** Cadmium 112	49 **In** Indium 115	50 **Sn** Tin 119	51 **Sb** Antimony 122	52 **Te** Tellurium 128	53 **I** Iodine 127	54 **Xe** Xenon 131
77 **Ir** Iridium 192	78 **Pt** Platinum 195	79 **Au** Gold 197	80 **Hg** Mercury 201	81 **Tl** Thallium 204	82 **Pb** Lead 207	83 **Bi** Bismuth 209	84 **Po** Polonium (209)	85 **At** Astatine (210)	86 **Rn** Radon (222)
109 **Une** Unnilennium (266)									

61 **Pm** Promethium (145)	62 **Sm** Samarium 150	63 **Eu** Europium 152	64 **Gd** Gadolinium 157	65 **Tb** Terbium 159	66 **Dy** Dysprosium 163	67 **Ho** Holmium 165	68 **Er** Erbium 167	69 **Tm** Thulium 169	70 **Yb** Ytterbium 173
93 **Np** Neptunium (237)	94 **Pu** Plutonium (244)	95 **Am** Americium (243)	96 **Cm** Curium (247)	97 **Bk** Berkelium (247)	98 **Cf** Californium (251)	99 **Es** Einsteinium (252)	100 **Fm** Fermium (257)	101 **Md** Mendelevium (258)	102 **No** Nobelium (259)

Understanding equations

As you read through this book, you will notice that many pages contain equations using symbols. If you are not familiar with these symbols, read this page. Symbols make it easy for chemists to write out the reactions that are occurring in a way that allows a better understanding of the processes involved.

Symbols for the elements

The basis of the modern use of symbols for elements dates back to the 19th century. At this time a shorthand was developed using the first letter of the element wherever possible. Thus "O" stands for oxygen, "H" stands for hydrogen

and so on. However, if we were to use only the first letter, then there could be some confusion. For example, nitrogen and nickel would both use the symbols N. To overcome this problem, many elements are symbolised using the first two letters of their full name, and the second letter is lowercase. Thus although nitrogen is N, nickel becomes Ni. Not all symbols come from the English name; many use the Latin name instead. This is why, for example, gold is not G but Au (for the Latin *aurum*) and sodium has the symbol Na, from the Latin *natrium*.

Compounds of elements are made by combining letters. Thus the molecule carbon

Written and symbolic equations

In this book, important chemical equations are briefly stated in words (these are called word equations), and are then shown in their symbolic form along with the states.

What reaction the equation illustrates

EQUATION: The formation of calcium hydroxide

Word equation ———— *Calcium oxide + water ⇨ calcium hydroxide*

Symbol equation ———— $CaO(s)$ + $H_2O(l)$ ⇨ $Ca(OH)_2(aq)$

heated

Sometimes you will find additional descriptions below the symbolic equation.

Symbol showing the state:
s is for solid, l is for liquid,
g is for gas and aq is for aqueous.

Diagrams

Some of the equations are shown as graphic representations.

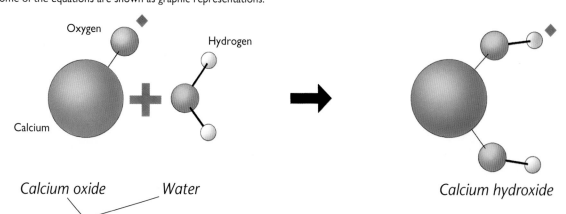

Oxygen

Hydrogen

Calcium

Calcium oxide   *Water*

Calcium hydroxide

Sometimes the written equation is broken up and put below the relevant stages in the graphic representation.

monoxide is CO. By using lowercase letters for the second letter of an element, it is possible to show that cobalt, symbol Co, is not the same as the molecule carbon monoxide, CO.

However, the letters can be made to do much more than this. In many molecules, atoms combine in unequal numbers. So, for example, carbon dioxide has one atom of carbon for every two of oxygen. This is shown by using the number 2 beside the oxygen, and the symbol becomes CO_2.

In practice, some groups of atoms combine as a unit with other substances. Thus, for example, calcium bicarbonate (one of the compounds used in some antacid pills) is written $Ca(HCO_3)_2$. This shows that the part of the substance inside the brackets reacts as a unit and the "2" outside the brackets shows the presence of two such units.

Some substances attract water molecules to themselves. To show this a dot is used. Thus the blue form of copper sulphate is written $CuSO_4.5H_2O$. In this case five molecules of water attract to one of copper sulphate.

When you see the dot, you know that this water can be driven off by heating; it is part of the crystal structure.

In a reaction substances change by rearranging the combinations of atoms. The way they change is shown by using the chemical symbols, placing those that will react (the starting materials, or reactants) on the left and the products of the reaction on the right. Between the two, chemists use an arrow to show which way the reaction is occurring.

It is possible to describe a reaction in words. This gives word equations, which are given throughout this book. However, it is easier to understand what is happening by using an equation containing symbols. These are also given in many places. They are not given when the equations are very complex.

In any equation both sides balance; that is, there must be an equal number of like atoms on both sides of the arrow. When you try to write down reactions, you, too, must balance your equation; you cannot have a few atoms left over at the end!

The symbols in brackets are abbreviations for the physical state of each substance taking part, so that (*s*) is used for solid, (*l*) for liquid, (*g*) for gas and (*aq*) for an aqueous solution, that is, a solution of a substance dissolved in water.

Atoms and ions
Each sphere represents a particle of an element. A particle can be an atom or an ion. Each atom or ion is associated with other atoms or ions through bonds – forces of attraction. The size of the particles and the nature of the bonds can be extremely important in determining the nature of the reaction or the properties of the compound.

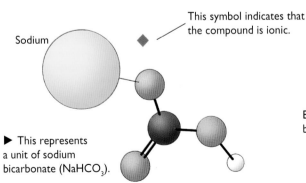

Sodium

This symbol indicates that the compound is ionic.

▶ This represents a unit of sodium bicarbonate ($NaHCO_3$).

The term "unit" is sometimes used to simplify the representation of a combination of ions.

Chemical symbols, equations and diagrams
The arrangement of any molecule or compound can be shown in one of the two ways shown below, depending on which gives the clearer picture. The left-hand diagram is called a ball-and-stick diagram because it uses rods and spheres to show the structure of the material. This example shows water, H_2O. There are two hydrogen atoms and one oxygen atom.

Bond shown by "stick"

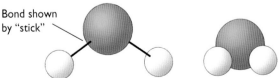

Colours too
The colours of each of the particles help differentiate the elements involved. The diagram can then be matched to the written and symbolic equation given with the diagram. In the case above, oxygen is red and hydrogen is grey.

Glossary of technical terms

absorb: to soak up a substance. Compare to adsorb.

acetone: a petroleum-based solvent.

acid: compounds containing hydrogen which can attack and dissolve many substances. Acids are described as weak or strong, dilute or concentrated, mineral or organic.

acidity: a general term for the strength of an acid in a solution.

acid rain: rain that is contaminated by acid gases such as sulphur dioxide and nitrogen oxides released by pollution.

adsorb/adsorption: to "collect" gas molecules or other particles on to the *surface* of a substance. They are not chemically combined and can be removed. (The process is called "adsorption".) Compare to absorb.

alchemy: the traditional "art" of working with chemicals that prevailed through the Middle Ages. One of the main challenges of alchemy was to make gold from lead. Alchemy faded away as scientific chemistry was developed in the 17th century.

alkali: a base in solution.

alkaline: the opposite of acidic. Alkalis are bases that dissolve, and alkaline materials are called basic materials. Solutions of alkalis have a pH greater than 7.0 because they contain relatively few hydrogen ions.

alloy: a mixture of a metal and various other elements.

alpha particle: a stable combination of two protons and two neutrons, which is ejected from the nucleus of a radioactive atom as it decays. An alpha particle is also the nucleus of the atom of helium. If it captures two electrons it can become a neutral helium atom.

amalgam: a liquid alloy of mercury with another metal.

amino acid: amino acids are organic compounds that are the building blocks for the proteins in the body.

amorphous: a solid in which the atoms are not arranged regularly (i.e. "glassy"). Compare with crystalline.

amphoteric: a metal that will react with both acids and alkalis.

anhydrous: a substance from which water has been removed by heating. Many hydrated salts are crystalline. When they are heated and the water is driven off, the material changes to an anhydrous powder.

anion: a negatively charged atom or group of atoms.

anode: the negative terminal of a battery or the positive electrode of an electrolysis cell.

anodising: a process that uses the effect of electrolysis to make a surface corrosion-resistant.

antacid: a common name for any compound that reacts with stomach acid to neutralise it.

antioxidant: a substance that prevents oxidation of some other substance.

aqueous: a solid dissolved in water. Usually used as "aqueous solution".

atom: the smallest particle of an element.

atomic number: the number of electrons or the number of protons in an atom.

atomised: broken up into a very fine mist. The term is used in connection with sprays and engine fuel systems.

aurora: the "northern lights" and "southern lights" that show as coloured bands of light in the night sky at high latitudes. They are associated with the way cosmic rays interact with oxygen and nitrogen in the air.

basalt: an igneous rock with a low proportion of silica (usually below 55%). It has microscopically small crystals.

base: a compound that may be soapy to the touch and that can react with an acid in water to form a salt and water.

battery: a series of electrochemical cells.

bauxite: an ore of aluminium, of which about half is aluminium oxide.

becquerel: a unit of radiation equal to one nuclear disintegration per second.

beta particle: a form of radiation in which electrons are emitted from an atom as the nucleus breaks down.

bleach: a substance that removes stains from materials either by oxidising or reducing the staining compound.

boiling point: the temperature at which a liquid boils, changing from a liquid to a gas.

bond: chemical bonding is either a transfer or sharing of electrons by two or more atoms. There are a number of types of chemical bond, some very strong (such as covalent bonds), others weak (such as hydrogen bonds). Chemical bonds form because the linked molecule is more stable than the unlinked atoms from which it formed. For example, the hydrogen molecule (H_2) is more stable

than single atoms of hydrogen, which is why hydrogen gas is always found as molecules of two hydrogen atoms.

brass: a metal alloy principally of copper and zinc.

brazing: a form of soldering, in which brass is used as the joining metal.

brine: a solution of salt (sodium chloride) in water.

bronze: an alloy principally of copper and tin.

buffer: a chemistry term meaning a mixture of substances in solution that resists a change in the acidity or alkalinity of the solution.

capillary action: the tendency of a liquid to be sucked into small spaces, such as between objects and through narrow-pore tubes. The force to do this comes from surface tension.

catalyst: a substance that speeds up a chemical reaction but itself remains unaltered at the end of the reaction.

cathode: the positive terminal of a battery or the negative electrode of an electrolysis cell.

cathodic protection: the technique of making the object that is to be protected from corrosion into the cathode of a cell. For example, a material, such as steel, is protected by coupling it with a more reactive metal, such as magnesium. Steel forms the cathode and magnesium the anode. Zinc protects steel in the same way.

cation: a positively charged atom or group of atoms.

caustic: a substance that can cause burns if it touches the skin.

cell: a vessel containing two electrodes and an electrolyte that can act as an electrical conductor.

ceramic: a material based on clay minerals, which has been heated so that it has chemically hardened.

chalk: a pure form of calcium carbonate made of the crushed bodies of microscopic sea creatures, such as plankton and algae.

change of state: a change between one of the three states of matter, solid, liquid and gas.

chlorination: adding chlorine to a substance.

cladding: a surface sheet of material designed to protect other materials from corrosion.

clay: a microscopically small plate-like mineral that makes up the bulk of many soils. It has a sticky feel when wet.

combustion: the special case of oxidisation of a substance where a considerable amount of heat and usually light are given out. Combustion is often referred to as "burning".

compound: a chemical consisting of two or more elements chemically bonded together. Calcium atoms can combine with carbon atoms and oxygen atoms to make calcium carbonate, a compound of all three atoms.

condensation nuclei: microscopic particles of dust, salt and other materials suspended in the air, which attract water molecules.

conduction: (i) the exchange of heat (heat conduction) by contact with another object or (ii) allowing the flow of electrons (electrical conduction).

convection: the exchange of heat energy with the surroundings produced by the flow of a fluid due to being heated or cooled.

corrosion: the *slow* decay of a substance resulting from contact with gases and liquids in the environment. The term is often applied to metals. Rust is the corrosion of iron.

corrosive: a substance, either an acid or an alkali, that *rapidly* attacks a wide range of other substances.

cosmic rays: particles that fly through space and bombard all atoms on the Earth's surface. When they interact with the atmosphere they produce showers of secondary particles.

covalent bond: the most common form of strong chemical bonding, which occurs when two atoms *share* electrons.

cracking: breaking down complex molecules into simpler components. It is a term particularly used in oil refining.

crude oil: a chemical mixture of petroleum liquids. Crude oil forms the raw material for an oil refinery.

crystal: a substance that has grown freely so that it can develop external faces. Compare with crystalline, where the atoms are not free to form individual crystals and amorphous where the atoms are arranged irregularly.

crystalline: the organisation of atoms into a rigid "honeycomb-like" pattern without distinct crystal faces.

crystal systems: seven patterns or systems into which all of the world's crystals can be grouped. They are: cubic, hexagonal, rhombohedral, tetragonal, orthorhombic, monoclinic and triclinic.

cubic crystal system: groupings of crystals that look like cubes.

curie: a unit of radiation. The amount of radiation emitted by 1 g of radium each second. (The curie is equal to 37 billion becquerels.)

current: an electric current is produced by a flow of electrons through a conducting solid or ions through a conducting liquid.

decay (radioactive decay): the way that a radioactive element changes into another element because of loss of mass through radiation. For example uranium decays (changes) to lead.

decompose: to break down a substance (for example by heat or with the aid of a catalyst) into simpler components. In such a chemical reaction only one substance is involved.

dehydration: the removal of water from a substance by heating it, placing it in a dry atmosphere, or through the action of a drying agent.

density: the mass per unit volume (e.g. g/cc).

desertification: a process whereby a soil is allowed to become degraded to a state in which crops can no longer grow, i.e. desert-like. Chemical desertification is usually the result of contamination with halides because of poor irrigation practices.

detergent: a petroleum-based chemical that removes dirt.

diaphragm: a semipermeable membrane – a kind of ultra-fine mesh filter – that will allow only small ions to pass through. It is used in the electrolysis of brine.

diffusion: the slow mixing of one substance with another until the two substances are evenly mixed.

digestive tract: the system of the body that forms the pathway for food and its waste products. It begins at the mouth and includes the stomach and the intestines.

dilute acid: an acid whose concentration has been reduced by a large proportion of water.

diode: a semiconducting device that allows an electric current to flow in only one direction.

disinfectant: a chemical that kills bacteria and other microorganisms.

dissociate: to break apart. In the case of acids it means to break up forming hydrogen ions. This is an example of ionisation. Strong acids dissociate completely. Weak acids are not completely ionised and a solution of a weak acid has a relatively low concentration of hydrogen ions.

dissolve: to break down a substance in a solution without a resultant reaction.

distillation: the process of separating mixtures by condensing the vapours through cooling.

doping: adding metal atoms to a region of silicon to make it semiconducting.

dye: a coloured substance that will stick to another substance, so that both appear coloured.

electrode: a conductor that forms one terminal of a cell.

electrolysis: an electrical–chemical process that uses an electric current to cause the break up of a compound and the movement of metal ions in a solution. The process happens in many natural situations (as for example in rusting) and is also commonly used in industry for purifying (refining) metals or for plating metal objects with a fine, even metal coating.

electrolyte: a solution that conducts electricity.

electron: a tiny, negatively charged particle that is part of an atom. The flow of electrons through a solid material such as a wire produces an electric current.

electroplating: depositing a thin layer of a metal onto the surface of another substance using electrolysis.

element: a substance that cannot be decomposed into simpler substances by chemical means

emulsion: tiny droplets of one substance dispersed in another. A common oil in water emulsion is milk. The tiny droplets in an emulsion tend to come together, so another stabilising substance is often needed to wrap the particles of grease and oil in a stable coat. Soaps and detergents are such agents. Photographic film is an example of a solid emulsion.

endothermic reaction: a reaction that takes heat from the surroundings. The reaction of carbon monoxide with a metal oxide is an example.

enzyme: organic catalysts in the form of proteins in the body that speed up chemical reactions. Every living cell contains hundreds of enzymes, which ensure that the processes of life continue. Should enzymes be made inoperative, such as through mercury poisoning, then death follows.

ester: organic compounds, formed by the reaction of an alcohol with an acid, which often have a fruity taste.

evaporation: the change of state of a liquid to a gas. Evaporation happens below the boiling point and is used as a method of separating out the materials in a solution.

exothermic reaction: a reaction that gives heat to the surroundings. Many oxidation reactions, for example, give out heat.

explosive: a substance which, when a shock is applied to it, decomposes very rapidly, releasing a very large amount of heat and creating a large volume of gases as a shock wave.

extrusion: forming a shape by pushing it through a die. For example, toothpaste is extruded through the cap (die) of the toothpaste tube.

fallout: radioactive particles that reach the ground from radioactive materials in the atmosphere.

fat: semi-solid energy-rich compounds derived from plants or animals and which are made of carbon, hydrogen and oxygen. Scientists call these esters.

feldspar: a mineral consisting of sheets of aluminium silicate. This is the mineral from which the clay in soils is made.

fertile: able to provide the nutrients needed for unrestricted plant growth.

filtration: the separation of a liquid from a solid using a membrane with small holes.

fission: the breakdown of the structure of an atom, popularly called "splitting the atom" because the atom is split into approximately two other nuclei. This is different from, for example, the small change that happens when radioactivity is emitted.

fixation of nitrogen: the processes that natural organisms, such as bacteria, use to turn the nitrogen of the air into ammonium compounds.

fixing: making solid and liquid nitrogen-containing compounds from nitrogen gas. The compounds that are formed can be used as fertilisers.

fluid: able to flow; either a liquid or a gas.

fluorescent: a substance that gives out visible light when struck by invisible waves such as ultraviolet rays.

flux: a material used to make it easier for a liquid to flow. A flux dissolves metal oxides and so prevents a metal from oxidising while being heated.

foam: a substance that is sufficiently gelatinous to be able to contain bubbles of gas. The gas bulks up the substance, making it behave as though it were semi-rigid.

fossil fuels: hydrocarbon compounds that have been formed from buried plant and animal remains. High pressures and temperatures lasting over millions of years are required. The fossil fuels are coal, oil and natural gas.

fraction: a group of similar components of a mixture. In the petroleum industry the light fractions of crude oil are those with the smallest molecules, while the medium and heavy fractions have larger molecules.

free radical: a very reactive atom or group with a "spare" electron.

freezing point: the temperature at which a substance changes from a liquid to a solid. It is the same temperature as the melting point.

fuel: a concentrated form of chemical energy. The main sources of fuels (called fossil fuels because they were formed by geological processes) are coal, crude oil and natural gas. Products include methane, propane and gasoline. The fuel for stars and space vehicles is hydrogen.

fuel rods: rods of uranium or other radioactive material used as a fuel in nuclear power stations.

fuming: an unstable liquid that gives off a gas. Very concentrated acid solutions are often fuming solutions.

fungicide: any chemical that is designed to kill fungi and control the spread of fungal spores.

fusion: combining atoms to form a heavier atom.

galvanising: applying a thin zinc coating to protect another metal.

gamma rays: waves of radiation produced as the nucleus of a radioactive element rearranges itself into a tighter cluster of protons and neutrons. Gamma rays carry enough energy to damage living cells.

gangue: the unwanted material in an ore.

gas: a form of matter in which the molecules form no definite shape and are free to move about to fill any vessel they are put in.

gelatinous: a term meaning made with water. Because a gelatinous precipitate is mostly water, it is of a similar density to water and will float or lie suspended in the liquid.

gelling agent: a semi-solid jelly-like substance.

gemstone: a wide range of minerals valued by people, both as crystals (such as emerald) and as decorative stones (such as agate). There is no single chemical formula for a gemstone.

glass: a transparent silicate without any crystal growth. It has a glassy lustre and breaks with a curved fracture. Note that some minerals have all these features and are therefore natural glasses. Household glass is a synthetic silicate.

glucose: the most common of the natural sugars. It occurs as the polymer known as cellulose, the fibre in plants. Starch is also a form of glucose. The breakdown of glucose provides the energy that animals need for life.

granite: an igneous rock with a high proportion of silica (usually over 65%). It has well-developed large crystals. The largest pink, grey or white crystals are feldspar.

Greenhouse Effect: an increase of the global air temperature as a result of heat released from burning fossil fuels being absorbed by carbon dioxide in the atmosphere.

gypsum: the name for calcium sulphate. It is commonly found as Plaster of Paris and wallboards.

half-life: the time it takes for the radiation coming from a sample of a radioactive element to decrease by half.

halide: a salt of one of the halogens (fluorine, chlorine, bromine and iodine).

halite: the mineral made of sodium chloride.

halogen: one of a group of elements including chlorine, bromine, iodine and fluorine.

heat-producing: see exothermic reaction.

high explosive: a form of explosive that will only work when it receives a shock from another explosive. High explosives are much more powerful than ordinary explosives. Gunpowder is not a high explosive.

hydrate: a solid compound in crystalline form that contains molecular water. Hydrates commonly form when a solution of a soluble salt is evaporated. The water that forms part of a hydrate crystal is known as the "water of crystallization". It can usually be removed by heating, leaving an anhydrous salt.

hydration: the absorption of water by a substance. Hydrated materials are not "wet" but remain firm, apparently dry, solids. In some cases, hydration makes the substance change colour, in many other cases there is no colour change, simply a change in volume.

hydrocarbon: a compound in which only hydrogen and carbon atoms are present. Most fuels are hydrocarbons, as is the simple plastic polyethene (known as polythene).

hydrogen bond: a type of attractive force that holds one molecule to another. It is one of the weaker forms of intermolecular attractive force.

hydrothermal: a process in which hot water is involved. It is usually used in the context of rock formation because hot water and other fluids sent outwards from liquid magmas are important carriers of metals and the minerals that form gemstones.

igneous rock: a rock that has solidified from molten rock, either volcanic lava on the Earth's surface or magma deep underground. In either case the rock develops a network of interlocking crystals.

incendiary: a substance designed to cause burning.

indicator: a substance or mixture of substances that change colour with acidity or alkalinity.

inert: nonreactive.

infra-red radiation: a form of light radiation where the wavelength of the waves is slightly longer than visible light. Most heat radiation is in the infra-red band.

insoluble: a substance that will not dissolve.

ion: an atom, or group of atoms, that has gained or lost one or more electrons and so developed an electrical charge. Ions behave differently from electrically neutral atoms and molecules. They can move in an electric field, and they can also bind strongly to solvent molecules such as water. Positively charged ions are called cations; negatively charged ions are called anions. Ions carry electrical current through solutions.

ionic bond: the form of bonding that occurs between two ions when the ions have opposite charges. Sodium cations bond with chloride anions to form common salt (NaCl) when a salty solution is evaporated. Ionic bonds are strong bonds except in the presence of a solvent.

ionise: to break up neutral molecules into oppositely charged ions or to convert atoms into ions by the loss of electrons.

ionisation: a process that creates ions.

irrigation: the application of water to fields to help plants grow during times when natural rainfall is sparse.

isotope: atoms that have the same number of protons in their nucleus, but which have different masses; for example, carbon-12 and carbon-14.

latent heat: the amount of heat that is absorbed or released during the process of changing state between gas, liquid or solid. For example, heat is absorbed when a substance melts and it is released again when the substance solidifies.

latex: (the Latin word for "liquid") a suspension of small polymer particles in water. The rubber that flows from a rubber tree is a natural latex. Some synthetic polymers are made as latexes, allowing polymerisation to take place in water.

lava: the material that flows from a volcano.

limestone: a form of calcium carbonate rock that is often formed of lime mud. Most limestones are light grey and have abundant fossils.

liquid: a form of matter that has a fixed volume but no fixed shape.

lode: a deposit in which a number of veins of a metal found close together.

lustre: the shininess of a substance.

magma: the molten rock that forms a balloon-shaped chamber in the rock below a volcano. It is fed by rock moving upwards from below the crust.

marble: a form of limestone that has been "baked" while deep inside mountains. This has caused the limestone to melt and reform into small interlocking crystals, making marble harder than limestone.

mass: the amount of matter in an object. In everyday use, the word weight is often used to mean mass.

melting point: the temperature at which a substance changes state from a solid to a liquid. It is the same as freezing point.

membrane: a thin flexible sheet. A semipermeable membrane has microscopic holes of a size that will selectively allow some ions and molecules to pass through but hold others back. It thus acts as a kind of sieve.

meniscus: the curved surface of a liquid that forms when it rises in a small bore, or capillary tube. The meniscus is convex (bulges upwards) for mercury and is concave (sags downwards) for water.

metal: a substance with a lustre, the ability to conduct heat and electricity and which is not brittle.

metallic bonding: a kind of bonding in which atoms reside in a "sea" of mobile electrons. This type of bonding allows metals to be good conductors and means that they are not brittle

metamorphic rock: formed either from igneous or sedimentary rocks, by heat and or pressure. Metamorphic rocks form deep inside mountains during periods of mountain building. They result from the remelting of rocks during which process crystals are able to grow. Metamorphic rocks often show signs of banding and partial melting.

micronutrient: an element that the body requires in small amounts. Another term is trace element.

mineral: a solid substance made of just one element or chemical compound. Calcite is a mineral because it consists only of calcium carbonate, halite is a mineral because it contains only sodium chloride, quartz is a mineral because it consists of only silicon dioxide.

mineral acid: an acid that does not contain carbon and that attacks minerals. Hydrochloric, sulphuric and nitric acids are the main mineral acids.

mineral-laden: a solution close to saturation.

mixture: a material that can be separated out into two or more substances using physical means.

molecule: a group of two or more atoms held together by chemical bonds.

monoclinic system: a grouping of crystals that look like double-ended chisel blades.

monomer: a building block of a larger chain molecule ("mono" means one, "mer" means part).

mordant: any chemical that allows dyes to stick to other substances.

native metal: a pure form of a metal, not combined as a compound. Native metal is more common in poorly reactive elements than in those that are very reactive.

neutralisation: the reaction of acids and bases to produce a salt and water. The reaction causes hydrogen from the acid and hydroxide from the base to be changed to water. For

example, hydrochloric acid reacts with sodium hydroxide to form common salt and water. The term is more generally used for any reaction where the pH changes towards 7.0, which is the pH of a neutral solution.

neutron: a particle inside the nucleus of an atom that is neutral and has no charge.

noncombustible: a substance that will not burn.

noble metal: silver, gold, platinum, and mercury. These are the least reactive metals.

nuclear energy: the heat energy produced as part of the changes that take place in the core, or nucleus, of an element's atoms.

nuclear reactions: reactions that occur in the core, or nucleus of an atom.

nutrients: soluble ions that are essential to life.

octane: one of the substances contained in fuel.

ore: a rock containing enough of a useful substance to make mining it worthwhile.

organic acid: an acid containing carbon and hydrogen.

organic substance: a substance that contains carbon.

osmosis: a process where molecules of a liquid solvent move through a membrane (filter) from a region of low concentration to a region of high concentration of solute.

oxidation: a reaction in which the oxidising agent removes electrons. (Note that oxidising agents do not have to contain oxygen.)

oxide: a compound that includes oxygen and one other element.

oxidise: the process of gaining oxygen. This can be part of a controlled chemical reaction, or it can be the result of exposing a substance to the air, where oxidation (a form of corrosion) will occur slowly, perhaps over months or years.

oxidising agent: a substance that removes electrons from another substance (and therefore is itself reduced).

ozone: a form of oxygen whose molecules contain three atoms of oxygen. Ozone is regarded as a beneficial gas when high in the atmosphere because it blocks ultraviolet rays. It is a harmful gas when breathed in, so low level ozone, which is produced as part of city smog, is regarded as a form of pollution. The ozone layer is the uppermost part of the stratosphere.

pan: the name given to a shallow pond of liquid. Pans are mainly used for separating solutions by evaporation.

patina: a surface coating that develops on metals and protects them from further corrosion.

percolate: to move slowly through the pores of a rock.

period: a row in the Periodic Table.

Periodic Table: a chart organising elements by atomic number and chemical properties into groups and periods.

pesticide: any chemical that is designed to control pests (unwanted organisms) that are harmful to plants or animals.

petroleum: a natural mixture of a range of gases, liquids and solids derived from the decomposed remains of plants and animals.

pH: a measure of the hydrogen ion concentration in a liquid. Neutral is pH 7.0; numbers greater than this are alkaline, smaller numbers are acidic.

phosphor: any material that glows when energized by ultraviolet or electron beams such as in fluorescent tubes and cathode ray tubes. Phosphors, such as phosphorus, emit light after the source of excitation is cut off. This is why they glow in the dark. By contrast, fluorescors, such as fluorite, emit light only while they are being excited by ultraviolet light or an electron beam.

photon: a parcel of light energy.

photosynthesis: the process by which plants use the energy of the Sun to make the compounds they need for life. In photosynthesis, six molecules of carbon dioxide from the air combine with six molecules of water, forming one molecule of glucose (sugar) and releasing six molecules of oxygen back into the atmosphere.

pigment: any solid material used to give a liquid a colour.

placer deposit: a kind of ore body made of a sediment that contains fragments of gold ore eroded from a mother lode and transported by rivers and/or ocean currents.

plastic (material): a carbon-based material consisting of long chains (polymers) of simple molecules. The word plastic is commonly restricted to synthetic polymers.

plastic (property): a material is plastic if it can be made to change shape easily. Plastic materials will remain in the new shape. (Compare with elastic, a property where a material goes back to its original shape.)

plating: adding a thin coat of one material to another to make it resistant to corrosion.

playa: a dried-up lake bed that is covered with salt deposits. From the Spanish word for beach.

poison gas: a form of gas that is used intentionally to produce widespread injury and death. (Many gases are poisonous, which is why many chemical reactions are performed in laboratory fume chambers, but they are a byproduct of a reaction and not intended to cause harm.)

polymer: a compound that is made of long chains by combining molecules (called monomers) as repeating units. ("Poly" means many, "mer" means part).

polymerisation: a chemical reaction in which large numbers of similar molecules arrange themselves into large molecules, usually long chains. This process usually happens when there is a suitable catalyst present. For example, ethene reacts to form polythene in the presence of certain catalysts.

porous: a material containing many small holes or cracks. Quite often the pores are connected, and liquids, such as water or oil, can move through them.

precious metal: silver, gold, platinum, iridium, and palladium. Each is prized for its rarity. This category is the equivalent of precious stones, or gemstones, for minerals.

precipitate: tiny solid particles formed as a result of a chemical reaction between two liquids or gases.

preservative: a substance that prevents the natural organic decay processes from occurring. Many substances can be used safely for this purpose, including sulphites and nitrogen gas.

product: a substance produced by a chemical reaction.

protein: molecules that help to build tissue and bone and therefore make new body cells. Proteins contain amino acids.

proton: a positively charged particle in the nucleus of an atom that balances out the charge of the surrounding electrons

pyrite: "mineral of fire". This name comes from the fact that pyrite (iron sulphide) will give off sparks if struck with a stone.

pyrometallurgy: refining a metal from its ore using heat. A blast furnace or smelter is the main equipment used.

radiation: the exchange of energy with the surroundings through the transmission of waves or particles of energy. Radiation is a form of energy transfer that can happen through space; no intervening medium is required (as would be the case for conduction and convection).

radioactive: a material that emits radiation or particles from the nucleus of its atoms.

radioactive decay: a change in a radioactive element due to loss of mass through radiation. For example uranium decays (changes) to lead.

radioisotope: a shortened version of the phrase radioactive isotope.

radiotracer: a radioactive isotope that is added to a stable, nonradioactive material in order to trace how it moves and its concentration.

reaction: the recombination of two substances using parts of each substance to produce new substances.

reactivity: the tendency of a substance to react with other substances. The term is most widely used in comparing the reactivity of metals. Metals are arranged in a reactivity series.

reagent: a starting material for a reaction.

recycling: the reuse of a material to save the time and energy required to extract new material from the Earth and to conserve non-renewable resources.

redox reaction: a reaction that involves reduction and oxidation.

reducing agent: a substance that gives electrons to another substance. Carbon monoxide is a reducing agent when passed over copper oxide, turning it to copper and producing carbon dioxide gas. Similarly, iron oxide is reduced to iron in a blast furnace. Sulphur dioxide is a reducing agent, used for bleaching bread.

reduction: the removal of oxygen from a substance. See also: oxidation.

refining: separating a mixture into the simpler substances of which it is made. In the case of a rock, it means the extraction of the metal that is mixed up in the rock. In the case of oil it means separating out the fractions of which it is made.

refractive index: the property of a transparent material that controls the angle at which total internal reflection will occur. The greater the refractive index, the more reflective the material will be.

resin: natural or synthetic polymers that can be moulded into solid objects or spun into thread.

rust: the corrosion of iron and steel.

saline: a solution in which most of the dissolved matter is sodium chloride (common salt).

salinisation: the concentration of salts, especially sodium chloride, in the upper layers of a soil due to poor methods of irrigation.

salts: compounds, often involving a metal, that are the reaction products of acids and bases. (Note "salt" is also the common word for sodium chloride, common salt or table salt.)

saponification: the term for a reaction between a fat and a base that produces a soap.

saturated: a state where a liquid can hold no more of a substance. If any more of the substance is added, it will not dissolve.

saturated solution: a solution that holds the maximum possible amount of dissolved material. The amount of material in solution varies with the temperature; cold solutions can hold less dissolved solid material than hot solutions. Gases are more soluble in cold liquids than hot liquids.

sediment: material that settles out at the bottom of a liquid when it is still.

semiconductor: a material of intermediate conductivity. Semiconductor devices often use silicon when they are made as part of diodes, transistors or integrated circuits.

semipermeable membrane: a thin (membrane) of material that acts as a fine sieve, allowing small molecules to pass, but holding large molecules back.

silicate: a compound containing silicon and oxygen (known as silica).

sintering: a process that happens at moderately high temperatures in some compounds. Grains begin to fuse together even through they do not melt. The most widespread example of sintering happens during the firing of clays to make ceramics.

slag: a mixture of substances that are waste products of a furnace. Most slags are composed mainly of silicates.

smelting: roasting a substance in order to extract the metal contained in it.

smog: a mixture of smoke and fog. The term is used to describe city fogs in which there is a large proportion of particulate matter (tiny pieces of carbon from exhausts) and also a high concentration of sulphur and nitrogen gases and probably ozone.

soldering: joining together two pieces of metal using solder, an alloy with a low melting point.

solid: a form of matter where a substance has a definite shape.

soluble: a substance that will readily dissolve in a solvent.

solute: the substance that dissolves in a solution (e.g. sodium chloride in salt water).

solution: a mixture of a liquid and at least one other substance (e.g. salt water). Mixtures can be separated out by physical means, for example by evaporation and cooling.

solvent: the main substance in a solution (e.g. water in salt water).

spontaneous combustion: the effect of a very reactive material beginning to oxidise very quickly and bursting into flame.

stable: able to exist without changing into another substance.

stratosphere: the part of the Earth's atmosphere that lies immediately above the region in which clouds form. It occurs between 12 and 50 km above the Earth's surface.

strong acid: an acid that has completely dissociated (ionised) in water. Mineral acids are strong acids.

sublimation: the change of a substance from solid to gas, or vica versa, without going through a liquid phase.

substance: a type of material, including mixtures.

sulphate: a compound that includes sulphur and oxygen, for example, calcium sulphate or gypsum.

sulphide: a sulphur compound that contains no oxygen.

sulphite: a sulphur compound that contains less oxygen than a sulphate.

surface tension: the force that operates on the surface of a liquid, which makes it act as though it were covered with an invisible elastic film.

suspension: tiny particles suspended in a liquid.

synthetic: does not occur naturally, but has to be manufactured.

tarnish: a coating that develops as a result of the reaction between a metal and substances in the air. The most common form of tarnishing is a very thin transparent oxide coating.

thermonuclear reactions: reactions that occur within atoms due to fusion, releasing an immensely concentrated amount of energy.

thermoplastic: a plastic that will soften, can repeatedly be moulded it into shape on heating and will set into the moulded shape as it cools.

thermoset: a plastic that will set into a moulded shape as it cools, but which cannot be made soft by reheating.

titration: a process of dripping one liquid into another in order to find out the amount needed to cause a neutral solution. An indicator is used to signal change.

toxic: poisonous enough to cause death.

translucent: almost transparent.

transmutation: the change of one element into another.

vapour: the gaseous form of a substance that is normally a liquid. For example, water vapour is the gaseous form of liquid water.

vein: a mineral deposit different from, and usually cutting across, the surrounding rocks. Most mineral and metal-bearing veins are deposits filling fractures. The veins were filled by hot, mineral-rich waters rising upwards from liquid volcanic magma. They are important sources of many metals, such as silver and gold, and also minerals such as gemstones. Veins are usually narrow, and were best suited to hand-mining. They are less exploited in the modern machine age.

viscous: slow moving, syrupy. A liquid that has a low viscosity is said to be mobile.

vitreous: glass-like.

volatile: readily forms a gas.

vulcanisation: forming cross-links between polymer chains to increase the strength of the whole polymer. Rubbers are vulcanised using sulphur when making tyres and other strong materials.

weak acid: an acid that has only partly dissociated (ionised) in water. Most organic acids are weak acids.·

weather: a term used by Earth scientists and derived from "weathering", meaning to react with water and gases of the environment.

weathering: the slow natural processes that break down rocks and reduce them to small fragments either by mechanical or chemical means.

welding: fusing two pieces of metal together using heat.

X-rays: a form of very short wave radiation.

Index